why you should store your farts in a jar

Also by David Haviland

Why Dogs Eat Poop and Other Useless or Gross
Information About the Animal Kingdom
(with Francesca Gould)

Other Books in the Series

Why You Shouldn't Eat Your Boogers and Other
Useless or Gross Information About Your Body
(by Francesca Gould)

Why Fish Fart and Other Useless or Gross
Information About the World
(by Francesca Gould)

why you should store your farts in a jar

and Other Oddball or Gross Maladies, Afflictions, Remedies, or "Cures"

david haviland

JEREMY P. TARCHER/PENGUIN

a member of Penguin Group (USA)

New York

JEREMY P. TARCHER/PENGUIN
Published by the Penguin Group
Penguin Group (USA) Inc., 375 Hudson Street, New York, New York
10014, USA • Penguin Group (Canada), 90 Eglinton Avenue East, Suite 700,
Toronto, Ontario M4P 2Y3, Canada (a division of Pearson Penguin
Canada Inc.) • Penguin Books Ltd, 80 Strand, London WC2R 0RL,
England • Penguin Ireland, 25 St Stephen's Green, Dublin 2, Ireland
(a division of Penguin Books Ltd) • Penguin Group (Australia), 250 Camberwell
Road, Camberwell, Victoria 3124, Australia (a division of Pearson Australia
Group Pty Ltd) • Penguin Books India Pvt Ltd,
11 Community Centre, Panchsheel Park, New Delhi–110 017, India • Penguin
Group (NZ), 67 Apollo Drive, Rosedale, North Shore 0632, New Zealand (a division
of Pearson New Zealand Ltd) • Penguin Books (South Africa) (Pty) Ltd,
24 Sturdee Avenue, Rosebank, Johannesburg 2196, South Africa

Penguin Books Ltd, Registered Offices: 80 Strand, London WC2R 0RL, England

The Hippocratic oath is quoted from *Hippocratic Writings*,
translated by J. Chadwick and W. N. Mann (Penguin, 1950).

Most Tarcher/Penguin books are available at special quantity discounts
for bulk purchase for sales promotions, premiums, fund-raising, and
educational needs. Special books or book excerpts also can be created
to fit specific needs. For details, write Penguin Group (USA) Inc.
Special Markets, 375 Hudson Street, New York, NY 10014.

Library of Congress Cataloging-in-Publication Data

Haviland, David.
Why you should store your farts in a jar : and other oddball or gross maladies,
afflictions, remedies, or "cures" / David Haviland.
p.   cm.
ISBN 978-1-58542-857-1
1. Medicine—Anecdotes.   2. Medicine—History—Anecdotes.   I. Title.
R705.H38   2010   2010029421
610—dc22

Printed in the United States of America
1 3 5 7 9 10 8 6 4 2

BOOK DESIGN BY NICOLE LAROCHE

For Beth

contents

chapter one

the wisdom of
the ancients

Physicians of all men are most happy;
what good success soever they have
the world proclaimeth,
and what faults they commit,
the earth covereth.

FRANCIS QUARLES

what unusual use did the ancient egyptians find for crocodile dung?

Bizarrely, crocodile dung was used as a contraceptive device in ancient Egypt. While the ancient Egyptians had a fairly complex and advanced system of medicine involving herbal drugs, poultices, laxatives, suppositories, surgery, bone setting, ophthalmology, and even a system of health insurance, the vast majority of Egyptian treatments were totally ineffective, and sometimes even quite harmful. For example, one Egyptian cure for impotence contained thirty-nine separate exotic ingredients, none of which would have had any useful effect. On the other hand, while crocodile dung as a method of contraception may sound pretty daft, it was probably effective to some degree. So how was it used? The dried dung was used as a pessary that was inserted into the vagina. The idea was that it would soften as it reached body temperature, and thus form a secure, impenetrable barrier against the cervix. Cervical caps that function in a similar way are used as contraception today, although thankfully they tend not to be made of dung. Furthermore, the acidity of the crocodile dung would probably have acted as a mild spermicide, offering some additional protection.

Nonetheless, putting crocodile dung into your body in any form is not recommended. Dung is full of bacteria, parasites, and other germs, so there is a considerable risk of infection. It's also just plain gross, not to mention stinky. Other traditional pessaries down through the ages have been made from elephant dung, tree sap, lemon halves, cotton, wool, and natural sea sponges, and each of them was probably only about as effective as the others.

what is the "doctrine of signatures"?

The "doctrine of signatures" is an ancient diagnostic system that is based upon the simple principle that God marked everything he created with a sign, and that this sign is a clue as to each thing's purpose. Therefore, walnuts were believed to be good for the brain, because they look a bit like brains, as well as being, like brains, a fairly soft, spongy substance enclosed within a hard shell. The doctrine of signatures was a central plank of medical theory going back at least as far as Galen of Pergamum (A.D. 129–200), the preeminent doctor of the Roman period, and it remained a central plank of medical thinking until as recently as the late nineteenth century.

According to the doctrine of signatures, natural remedies display or somehow reflect the ailments they were designed to cure. Thus, lungwort was believed to be good for lung conditions, because its oval, white-spotted leaves look a bit like diseased lungs. Cardamine flowers, also known as "tooth-wort," were meant to be good for toothache, as these small, white flowers have a shape similar to teeth. St. John's wort was said to be good for the skin, because the oil glands in the plant's leaves look like skin pores. In a few cases, there was some genuine truth to these claims—St. John's wort, for example, is recognized by modern medicine as containing a strong antibiotic that helps wounds heal quickly. (Incidentally, St. John's wort is also used in herbal medicine as an antidepressant, although recent studies are inconclusive as to whether or not it has anything more than a placebo effect.)

More often, though, the remedies recommended by the doctrine had no real medicinal value. For example, henbane was recommended for tooth-ache because its seed container is shaped like a human jaw, but henbane is actually a poisonous hallucinogen that is potentially fatal. Nowadays, of course, the doctrine of signatures is not taken seriously by mainstream medicine, and any successes it might claim are regarded as simply coincidences, or attribution after the fact. In other words, when people noticed that St. John's wort helped wounds heal, it would have made sense for them to find a way to compare it with skin, to fit in with their contemporary understanding of science.

However, there are two booming areas of modern health care in which the doctrine of signatures is still taken seriously. First, the doctrine is used by modern herbalists, either for clues to help them rediscover forgotten essences, or to support the claims made for existing herbal products. Second, the doctrine is one of the two defining principles of homeopathy, in which context it is often described as "like cures like." The other key homeopathic principle is the "law of infinitesimals," which holds that the smaller the dose, the more effective the medicine. In accordance with this second law, homeopathic remedies are diluted to such an enormous extent that they actually contain nothing but water, although advocates believe that this water has somehow retained a "memory" of the original substance. Despite being dismissed as quackery by mainstream science, homeopathy continues to be funded and supported by Britain's National Health Service.

which society believed dead mouse paste could cure toothache?

In ancient Egypt, one recommended cure for toothache was to apply a dead mouse to the tooth or gum. Alternatively, the patient could mash the mouse into a paste, and mix it with other ingredients before applying it.

The ancient Egyptians weren't alone in extolling the benefits of mouse poultices. In Elizabethan England, one cure for warts was to cut a mouse in half and apply it to the offending pustule. The Elizabethans also ate mice, either fried or baked in pies. As well as curing warts, mice were also believed to remedy whooping cough, measles, smallpox, and bed-wetting.

where do maggots come from?

Maggots are the larvae of flies. They emerge from eggs which have been laid directly onto the kinds of things that flies consider food, such as rotting animal corpses. For centuries, maggots were believed to simply emerge from the corpse, in a process called "spontaneous generation." In other words, people believed that living things could be generated from nonliving things, without the presence of an egg, larva, or parent. Spontaneous generation was the accepted explanation for the origins of many types of creature for over two thousand years.

The theory was notably advanced by Aristotle in his text *The History of Animals*. He believed fleas and maggots emerged spontaneously from putrefying flesh, mice were created by piles of dirty hay, and aphids would magically arise from the morning dew. Some creatures clearly were the product of parents, and of course Aristotle recognized this, but he thought there was also another class of creatures that could emerge from nonliving things. Astonishingly, this theory of spontaneous generation held sway until as late as the nineteenth century.

With the development of increasingly powerful microscopes, scientists began to observe smaller and smaller life-forms, which once again raised the question of generation: Were these tiny microorganisms proof of spontaneous generation, or did they show that there was a whole world of tiny life-forms that we knew nothing about? A Dutch scientist named Antonie van Leeuwenhoek observed tiny creatures in the microscopes he expertly built for himself, and he called these tiny life-forms "animalcules." Van Leeuwenhoek noticed that living microorganisms would appear in rainwater after just a few days, which raised the question of where they had come from.

In the eighteenth and nineteenth centuries, a series of experiments would conclusively answer this question, paving the way for modern, effective medicine. In 1748, John Needham performed a series of experiments by boiling jars of meat broth. Needham believed that boiling the broth would kill any living animalcules, and so when the broth turned cloudy soon afterward, indicating the presence of microorganisms, this confirmed his theory that the animalcules had spontaneously emerged from the broth. In 1765, Lazzaro Spallanzani carried out a similar experiment, but sealed the lids of the jars, making them airtight to prevent any contaminants from getting in after boiling. Because the jars were sealed, the broths did not grow any microbes. Critics argued that no air was allowed to reach the broth, and air was one of the elements that was believed to be crucial in somehow facilitating spontaneous generation.

In 1859, Louis Pasteur resolved this difficult impasse by boiling broth in a specially designed swan-necked flask, which had an extremely long, thin neck that curved downward. In this way, the neck would allow air to reach the broth, but no microorganisms could navigate the narrow bend. After boiling, the flask remained free of microbial growth, demonstrating that air alone wasn't sufficient to generate microorganisms. This elegant experiment, which could be easily repeated and tested by anyone, clearly disproved the theory of spontaneous generation, and was thus a key moment in the development of germ theory.

when was plastic surgery invented?

Surprisingly perhaps, plastic surgery has been around for more than three thousand years. In India, records have been found detailing ancient procedures for repairing a broken nose, and suturing to avoid scarring. Around 500 B.C., a Hindu doctor named Susruta developed a procedure to repair noses that had been cut off as a punishment for adultery (for some reason, it was determined that the interloper in the marriage was most at fault, so it was this third party who would lose his or her nose). Susruta found a way to repair this shameful injury, which was of course intended to cause the maximum social stigma, by taking skin from the cheek or forehead.

In 1597, Italian doctor Gasparo Tagliacozzi improved the procedure by lifting a skin flap from the arm and stitching it onto the nose, while it was still simultaneously attached to the arm. Once the graft had taken, the flap would be cut free from the arm. There was considerable demand for nose surgery in Europe from the fifteenth century onward, because of the dreadful effects of syphilis, which could cause sufferers to lose their nose.

Before the development of effective anesthetic in the 1840s, any kind of surgery was incredibly painful, not to mention dangerous, so the idea of plastic surgery for purely cosmetic reasons was unthinkable. Interestingly, the word *plastic* in the context of plastic surgery doesn't mean artificial, nor does it refer to the materials used. Instead, it comes from the Greek word *plastikos*, which means to mold or shape, in the same sense that ceramics and sculpture are known as the "plastic arts."

were the incas the first to carry out blood transfusions?

In the history of European medicine, the key development in the story of blood transfusions was Karl Landsteiner's discovery in 1901 that different people have different blood types. Before this point, blood transfusions had been tried many times, but they were often fatal, because our bodies undergo an immunological reaction to blood of a different type. Landsteiner correctly concluded that it was necessary to match the blood type of the donor to that of the recipient.

Before this discovery, many types of blood transfusion had been attempted. In 1492, Pope Innocent VIII was given the blood of three young boys after slipping into a coma, but he and the boys all died as a result. In 1667, French doctor Jean-Baptiste Denys gave 8 ounces (approx. 25 ml) of lambs' blood to a young man with a fever who had been repeatedly bled, and the man survived. Denys then performed a transfusion into a laborer, who also survived. In both cases the amount of blood transfused was relatively small, which presumably explains how the patients managed to survive the allergic reaction. Denys's next patient died, and as a result blood transfusions were banned in both France and Britain. However, rumors circulated that the unfortunate patient may have died not from the transfusion, but from poison administered by his own wife!

There are some reports which suggest that blood transfusions may have been successfully performed centuries before Denys, by the Inca people of South America. The Incan Empire rapidly expanded during the fifteenth and sixteenth centuries, covering much of the west coast of South America, from as far south as present-day Argentina to as far north as present-day Colombia. The best record of their civilization was written by "El Inca" Garcilaso de la Vega, who was the son of a Spanish conquistador and an Incan princess. Garcilaso described the Incan system of medicine as a bloody business, based largely around bloodletting and purging. Patients would be bled from the arm or leg nearest to the site of the ailment; a headache, for example, was treated by bleeding from the forehead, at the spot between the eyebrows.

The reason why the Incas may have been able to succeed where European medicine had until that point failed was simply that they may all have had the same blood type. It seems that a vast majority of Incas were of blood type O, rhesus positive, which means that there would have been very few incompatibility reactions.

does the hippocratic oath specify that doctors should not have sex with their patients?

Hippocrates (c. 460–377 B.C.) was born on the island of Cos, and is regarded as the father of modern medicine. He and his followers published a great number of medical texts, which formed the foundation of Western medicine until the Enlightenment. Hippocrates was the first doctor to reject the prevailing superstitious belief that illness was caused by the gods. Instead, Hippocrates argued that illness was the product of the patient's environment, diet, and lifestyle, and that therefore professional physicians could bring about natural means of healing, without requiring the intervention of the gods. The Hippocratic school left behind around sixty works, which are known as the Hippocratic *corpus*, that formed the basis of medicine until the nineteenth century.

To address concerns about the ethics of medical practice, Hippocrates and his followers produced a detailed oath to demonstrate the physician's devotion to his art and to the patient. It is in this oath that the first principle of Hippocratic medicine is outlined: to do no harm—"*primum non nocere.*" A version of the oath is still taken by the majority of doctors today. When people refer to the Hippocratic oath nowadays, they usually mean only the principle that anything told to the doctor must be treated as confidential, and it is true that this is one of the principles of the oath. It contains a number of other points. The following are a number of key lines from the oath:

> *I swear by Apollo the healer, by Aesculapius, by Health and all the powers of healing, and call to witness all the gods and goddesses that I may keep this Oath and Promise to the best of my ability and judgment.*
>
> *I will use my power to help the sick to the best of my ability and judgment; I will abstain from harming or wronging any man by it.*
>
> *I will not give a fatal draught to anyone if I am asked, nor will I suggest any such thing. Neither will I give a woman means to procure an abortion.*
>
> *I will not cut, even for the stone, but I will leave such procedures to the practitioners of that craft.*

> *Whenever I go into a house, I will go to help the sick and never with the intention of doing harm or injury. I will not abuse my position to indulge in sexual contacts with the bodies of women or of men, whether they be freemen or slaves.*
>
> *Whatever I see or hear, professionally or privately, which ought not to be divulged, I will keep secret and tell no one.*
>
> *If, therefore, I observe this Oath and do not violate it, may I prosper both in my life and in my profession, earning good repute among all men for my time. If I transgress and forswear this oath, may my lot be otherwise.*

The boldface lines do seem to state that takers of the oath will not perform euthanasia, abortion, or surgery, so how do modern-day doctors reconcile this oath with today's practices? Quite simply, they take a different oath. The most common oaths taken today are the declaration of Geneva, the prayer of Maimonides, the oath of Lasagna, and the reinstatement of the Hippocratic oath. All four are based on the Hippocratic oath, and contain many of the same vows, but each is amended to be more appropriate for modern ethics and medical practices.

Most graduating medical students today take one

of these four versions of the Hippocratic oath before going out into the world to practice medicine. In a survey of the oaths taken in 150 U.S. and Canadian medical schools in 1993, it was found that only 14 percent forbade euthanasia, 8 percent prohibited abortion, and only 3 percent banned sexual contact with patients. The line referring to surgery is usually retained in some form, but in today's oaths it is treated as a metaphor, for the purpose of acknowledging that no doctor can maintain expertise in all fields.

was julius caesar delivered via cesarean section?

There is an enduring myth that the Roman dictator Julius Caesar was born by cesarean section, and that consequently either the operation was named after him, or he was named after the operation. This story is widely known, and has even made it into the *Oxford English Dictionary*. Nonetheless, it is surely false, as we shall see.

First, what exactly is a cesarean section? It is an operation in which an expectant mother's abdomen is cut open to deliver the baby when a vaginal delivery is thought to be too dangerous. At the time of Julius Caesar's birth (and for centuries afterward) a cesarean birth was an extreme measure that would only be carried out if the mother had died or was likely to die. In Roman times, there was no effective anesthesia, and doctors had no proper way of suturing the enormous wounds that constitute the operation. Even on those rare occasions when doctors did manage to stem the hemorrhaging, the wounds were pretty much certain to become infected.

Cesarean births are known to have predated Julius Caesar, but there are no records of any mother surviving a cesarean birth until A.D. 1500. This is one reason we can be sure Caesar himself was not born by this method, as his mother didn't die until he was forty-six. Also, Julius Caesar couldn't have been named after the operation, because the name Caesar had already been in the family for generations. The dynasty kept detailed ancestral records, and claimed that they could trace their lineage back to the Trojan prince Aeneas and the goddess Venus.

Although Julius Caesar himself was not born by cesarean, there does seem to be a link between the two names. The name "Caesar" may have entered the family because one of Julius Caesar's distant ancestors was born by this method, according to the Roman writer Pliny the Elder. The name may therefore have come from the Latin word *caedere*, meaning "to cut," or from a law called Lex Caesarea, which required the operation to take place if the mother died during childbirth.

Alternatively, the family name may have had nothing to do with the cesarean section. Instead, it may have referred to an ancestor who was born with a full head of hair, from *caesaries*, meaning "mane" or "luxuriant hair"; or an ancestor who had blue-gray eyes, from *caesius*, meaning "blue-gray" or from *caesai*, the Moorish or Punic word for "elephant," in reference to an ancestor who killed an elephant during the First Punic War. During his reign, Julius Caesar had coins struck with an image of an elephant above the name Caesar, which suggests that this was his preferred theory.

what is "sounding"?

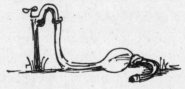

Before the development of antibiotics in the twentieth century, there was no effective treatment for gonorrhea, a common sexually transmitted disease that can cause sterility. Male sufferers find it painful to urinate, and may produce a thick yellow discharge, known as "gleet." Infected females may have similar symptoms, but are often symptomless. But if the infection is allowed to spread, it can cause severe complications.

Gonorrhea can cause scars and obstructions to grow inside the urethra, which is the tube that leads from the bladder to the genital opening; it is the tube through which we urinate. In men, it runs through the length of the penis. In women, it emerges just above the vaginal opening. If the urethra is blocked, it can make urination painful or even impossible. Sounding is not a cure for gonorrhea, it is merely a treatment to address these obstructions. The doctor inserts a long metal probe into the opening of the urethra, and uses this device, which is called a "sound," to stretch and widen the tube, clearing any blockage. It is said to be an extremely unpleasant experience, and men have been known to faint as a result.

Nowadays, gonorrhea can be treated with antibiotics, so why does sounding live on? One answer is that some kinky people find it rather fun. Fetish websites sell sets of sounds, of varying shapes and sizes, along with guides as to their use. One apparently pleasurable practice is to insert a curved sound down the urethra, deep into the bladder, to stimulate the male prostate. There are even piercings which incorporate sounds. One such is the Prince's Wand, a urethral sound held in place by a Prince Albert piercing, which as you may know is a thick ring which is inserted into the glans of the penis. I mean, really, isn't life painful enough?

chapter two

disgusting diseases

Where the sun enters, the doctor does not.

TRADITIONAL PROVERB

is it true that there are men who have to wheel their testicles around on a wheelbarrow?

Amazingly, some sufferers of elephantiasis really do have to do just this, because their scrotums have swollen to such an enormous size. Elephantiasis is an unpleasant, disfiguring disease that is found in many parts of Africa, India, and South Asia, and affects more than 120 million people worldwide. In communities where the disease is endemic, it can affect up to 50 percent of men, and around 10 percent of women.

Elephantiasis is caused by long, thread-like filarial worms, which are spread by mosquitoes and reproduce in the human bloodstream, infecting the lymphatic system. Infected people often remain symptomless for years, while the worms rapidly multiply in their blood. When symptoms do appear, they are dramatic. Sufferers develop grotesquely swollen limbs, genitals, and breasts. Men are more likely to be victims than women, and swelling of the penis and scrotum is common.

Recent developments have made elephantiasis easier to diagnose in symptomless sufferers, and the disease can be treated with antibiotics. Surgery can also be helpful for victims of scrotal elephantiasis. The World Health Organization is working toward an ambitious target to completely eliminate the disease by 2020, and so far it says it's on track to meet this goal.

so does this mean there will never be another elephant man?

Actually, although Joseph Merrick, the man who came to be known as the Elephant Man, was thought to have suffered from elephantiasis, it now seems more likely that he was in fact suffering from a combination of two rare disorders: Proteus syndrome and neurofibromatosis type I.

He was born in Leicester in 1862, with no visible signs of any unusual condition. Merrick was a healthy, normal child until around the age of five, when his skin began to change. It became thick, lumpy, and gray, and he developed a swelling on his upper lip. As he grew older, his right arm became grotesquely swollen, while his left arm remained thin but normal. The swellings on his face continued to grow, along with a large, bony lump on his forehead. At one point he fell and damaged his hip, leaving him permanently lame.

After leaving school, he took on a number of jobs, but his deteriorating condition and shocking appearance made work impossible. He worked in a factory rolling cigars, until his fingers became too thick and lost the quickness and dexterity required for the job. Astonishingly, he then worked as a door-to-door salesman, but people were shocked by his appearance, and the swellings on his face meant that his speech had become incomprehensible. He ended up in a workhouse before deciding that his only option was to join a freak show.

Merrick teamed up with a promoter, who exhibited him to the public as the "Elephant Man," along with a detailed biography explaining that Merrick's mother had been knocked down by a rampaging elephant during pregnancy, which had caused the baby to be born disfigured. It sounds like a showman's over-the-top invention, but in fact there was an elephant on the loose in Leicester in 1862, so there may be some truth to the story, although of course even if Merrick's mother had been knocked over by an elephant, it wouldn't have caused his condition.

One of the visitors to the freak show was a doctor named Frederick Treves, who took a particular interest in Merrick, and examined him at his hospital on a number of occasions. After a disastrous attempt to take the show to Europe, Merrick was robbed and left destitute, and so Treves admitted him to live at the London Hospital. After Treves organized a public campaign to raise funds for his care, Merrick became a figure who was looked on with a great deal of public sympathy, and was befriended by many of the great and good, including Princess Alexandra and Queen Victoria.

Apart from the exact nature of his condition, there remains another enduring mystery concerning

Merrick: How did he die? David Lynch's film *The Elephant Man* shows Merrick suffocating in his sleep, but this is not what is believed to have happened. When he was found, Merrick had dislocated his neck, and it was this which had killed him. Merrick's head was too heavy for him to sleep in a reclining position, but it's possible that on this occasion he may have wanted to forget his disfigurements for one night, and lay down to sleep like a normal person.

Despite the unimaginable hardships he faced, Merrick is said to have been a gentle and cultured man. He often ended his correspondence with the following poem by Isaac Watts:

> 'Tis true my form is something odd,
> But blaming me is blaming God.
> Could I create myself anew,
> I would not fail in pleasing you.
> If I could reach from pole to pole,
> Or grasp the ocean with a span,
> I would be measured by the soul,
> The mind's the standard of the man.

Today, there are effective treatments and surgical procedures for both Proteus syndrome and neurofibromatosis type I. Since the symptoms of someone like Joseph Merrick are so visibly dramatic, it seems unlikely that there will ever be another Elephant Man, as today anyone exhibiting such symptoms would surely receive quick and effective treatment.

how frequently can a person vomit in a twelve-hour period?

Probably more often than you might imagine. There is a particular condition called cyclic vomiting syndrome (CVS) in which sufferers regularly vomit as much as twelve times per hour—that's every five minutes—for weeks on end. In between these periods, there will be intervals with no symptoms, and there is usually a set pattern of episodes. This dreadful condition was first described in 1882 by Doctor Samuel Gee. It tends to develop in children between the age of three and seven, and it can continue into adulthood. Many sufferers find the condition too debilitating to go to school or work, while others seem to be less dramatically affected.

It's not clear what causes CVS, but it is thought to be somehow linked to migraine headaches. Many CVS sufferers also have a family history of migraine, and the two conditions are quite similar in some ways: They both consist of sudden, intense symptoms, some of which overlap, with pain-free intervals, and both are triggered by many of the same stimuli. Some CVS sufferers say that their triggers are very predictable, and they can include illness, stress, certain foods, excitement, anxiety, and panic attacks.

Rest and sleep are normally recommended when a CVS bout begins, and many sufferers learn to physically control their symptoms. Frequent vomiting can cause dehydration and a loss of electrolytes, so sufferers are sometimes given painkillers or antinausea medicine. Vomiting can also irritate or injure the esophagus because of the stomach acid, so vomit will sometimes contain blood and bile.

what was the "king's evil"?

This was the colloquial name given to a horrible disease called scrofula, which is a tuberculous infection of the skin of the neck. Scrofula affects the lymph nodes, causing large blue or purple swellings to grow on the neck and chest. These lumps are initially painless, if unsightly, but after a time, they burst and leave a nasty open wound.

Scrofula is usually caused by *Mycobacterium tuberculosis* in adults, which can be passed from person to person through breathing. As the prevalence of tuberculosis declined during the twentieth century, scrofula was almost wiped out, but with the advent of HIV-weakened immune systems, it is becoming more common again.

While today most types of scrofula can be easily treated with antibiotics, for centuries there was no effective treatment or cure. The disease became known as the "King's Evil," because it was thought that a touch from the monarch would cure the swellings. This royal healing power was said to have been passed down from Edward the Confessor. Many of the kings of England and France would indulge in this practice, touching great numbers of victims, and handing out coins to the afflicted. Henry IV of France is reported to have touched as many as 1,500 scrofula victims in one sitting. Charles II of England is recorded as bestowing the royal touch on 92,107 of his subjects over the course of his reign, although quite why such detailed records were kept is not clear. The great man of letters Dr. Johnson was touched by Queen Anne for scrofula when he was a child. The practice died out in the eighteenth century, as it was felt, in England at least, to be too Catholic. It may also have taken a blow in France when the Dauphin Louis XVII died of scrofula; this, you would think, undermined the tradition.

why did people believe that keeping farts in a jar could ward off the black death?

The Black Death was one of the deadliest plagues in human history. Between 1348 and 1350 it killed around 1.5 million people in Britain, out of a population of just 4 million. The plague was characterized by unpleasant black buboes that would form in the victims' groin, neck, and armpits, oozing pus and blood. It caused fever, nausea, and vomiting, and most victims died within four to seven days of becoming infected. There were regular outbreaks of plague about once every generation, until the last major outbreak in 1665, the Great Plague of London.

Medicine of this period was based upon principles that had been around since the ancient Greeks, including the simple tenet that "like cures like." Since it was believed that the plague was caused by deadly vapors, it therefore made sense that other foul smells might help to ward off the disease. Some doctors therefore recommended keeping dirty goats inside the home, to create a therapeutic stink. Others suggested using another source of foul odors: our own farts. However, rather than waste the precious pungencies, people were instead advised to store their farts in jars, which could then be opened and inhaled the next time the deadly pestilence appeared in the neighborhood.

which disease is called "trembling with fear"?

The answer is *kuru*, a fascinating, incurable brain disease that, so far at least, has only ever affected a tiny tribe known as the Fore (pronounced for-ay) people, who are found in the highlands of New Guinea. The word *kuru* means "trembling with fear" in the Fore's native language, and it refers both to a key symptom of the disease, namely physical tremors, and its dreadful fatality rate. After the onset of symptoms, kuru victims were almost certain to die within six to twenty-four months, making kuru one of the deadliest diseases known to man.

Kuru was first discovered in 1957, and at that point it may not have even existed for very long. According to the older members of the tribe, the disease had not been known when they were young, which suggests it couldn't have been around for much more than twenty years. The Fore people were completely isolated, thanks to the mountainous terrain of New Guinea, and there are no reports of kuru occurring anywhere else in the world.

The disease is believed to have spread as a result

of the Fore's cannibalistic funeral rites. When a member of the tribe died, the female relatives would ritualistically dismember the corpse, removing the arms and feet, stripping the muscle from the body, cutting open the chest to remove the internal organs, and scooping out the brain. The tribe would then cook and eat the corpse, including the brain, which is thought to be the most infectious organ.

Meat from the bodies of kuru victims was highly prized, as the layers of fat on those who had died quickly would apparently resemble pork. The men of the tribe were given the best cuts of meat, leaving the rest of the body, including the brain, to the women and children, which explains why kuru was around eight to nine times more prevalent among women than among men. An alternative or perhaps additional explanation for this fact is that it was the women who were responsible for dismembering the bodies, and they may therefore have become infected via open sores and cuts coming into contact with the disease, rather than because of ingesting the brain material.

The disease's effects were dramatic, and an epidemic quickly followed. The first symptoms of kuru were headaches, joint pain, physical tremors, and a gradual loss of limb control. Some victims also burst into pathological fits of laughter. Over time, sufferers became unable to stand, and then unable to eat. As a result, many died of starvation during the epidemic. Between 1957 and 1968, over 1,100 people died of kuru, out of a population of just 8,000. There is no cure or treatment, but the disease gradually died out, as the efforts of the Australian government and Christian missionaries eventually persuaded the Fore to discontinue their cannibalistic death rituals.

Kuru was an extraordinary disease in one other respect. It was an infectious disease, but it was not caused by a virus, bacterium, or a parasite. Instead, it was caused by prions, which are simply misshapen proteins, which somehow cause other proteins in the body to lose their shape. Kuru was similar to other prion diseases, including Creutzfeldt-Jakob disease, BSE (mad cow disease), and scrapie, collectively known as transmissible spongiform encephalopathies (TSEs), which essentially means that they make the victim's brain spongy and full of holes. Thankfully, kuru now seems to have been completely eliminated.

why would the ebola virus be a poor choice for biological warfare?

Ebola is a horrific disease that destroys blood vessel cells, leading to massive internal bleeding. It is infectious, highly fatal, and endemic in parts of tropical Africa. The disease was first observed in Zaire in

1976 in an outbreak that infected more than three hundred people, 90 percent of whom died.

Although the Ebola virus is both deadly and contagious, it would actually make a poor choice as a biological weapon for one simple reason: It kills its victims too quickly. Viruses such as influenza can spread quickly over whole continents because they are easily transmitted, their initial symptoms are mild, and their victims are often contagious for a relatively long time.

Ebola, on the other hand, is not so easily transmitted, for a number of reasons. First, it does not seem to be particularly contagious in its early stages. After an incubation period of around five to eighteen days, symptoms will begin, and these can be severe, including fever, abdominal pain, and bloody vomit. After the onset of these symptoms, most patients have only around two weeks to live—most die of multiple organ failure or massive loss of blood. The "case fatality rate" of some strains of Ebola may be as high as 90 percent, meaning that 90 percent of those infected die from the disease. By comparison, the case fatality rate for Spanish flu, the worst pandemic in human history, which killed around 50 million people between 1918 and 1920, was only around 5 to 10 percent. Ebola victims are contagious for such a short time, with such severe symptoms, that there is only limited scope for them to transmit the disease to others.

While Ebola does not seem to be an ideal choice

for biological warfare, this doesn't mean people haven't tried. In 1992, a bizarre Japanese religious cult called Aum Shinrikyo tried to acquire a sample of the Ebola virus for use as a biological weapon. Cult leader Shoko Asahara led a group of around forty followers to Zaire, under the pretext of a medical mission, to try and get hold of a sample of the disease. The group were later convicted of the Sarin gas attack on the Tokyo subway in 1995, and when police raided the cult headquarters at the foot of Mount Fuji, they found a range of chemical and biological warfare agents, including anthrax and Ebola, as well as guns, explosives, and a Russian Mil Mi-17 military helicopter. The Sarin gas attack was the worst terrorist atrocity in Japanese history, and the group's motives remain unclear. One theory is that the attack was, bizarrely, an attempt to draw police attention away from an ongoing investigation into the group.

can people grow horns?

Amazingly, they can! There is a rare condition called *cornu cutaneum*, in which sufferers grow strange, conical protrusions that are hard and brittle, resembling horn, wood, or coral. These horns, which are known as cutaneous horns, are usually found on visible parts of the body, such as the face, ears, forearms, and hands, which suggests that they may be linked to radiation from sun exposure, as these are the parts of the body that tend to see the sun. They may also be linked to the human papilloma virus family, which causes warts.

Cutaneous horns are tumors that can be benign, premalignant, or malignant, and surgery is usually the recommended treatment. They are made of keratin, which is a versatile protein found in our hair and fingernails, as well as in various animals' horns, hooves, and claws.

In one astonishing recent case, an elderly Chinese woman was found who appeared to be growing large, devil-like horns. On one side of her forehead, the 101-year-old Zhang Ruifang had grown a thick, black horn that was more than 2¼ inches (6 cm) long. Then, another matching horn started to grow on the other side of her forehead, making her look quite, well, devilish. . . .

do we share any diseases with animals?

Indeed we do. In fact, we share hundreds of diseases with animals, and they are known as *zoonoses* (the singular is zoonosis). A great many of the diseases that afflict humanity today were first suffered by animals. At the end of the last Ice Age, around ten to twelve thousand years ago, humans began to move from a nomadic hunter-gatherer society to one based on settlements and agriculture. One consequence of this was that we began to live in close proximity with animals, including poultry, dogs, pigs, horses, sheep, and cattle.

At this point, illnesses that had previously evolved to infect animals began the process of mutating and leaping the species gap to infect humans. From horses, we caught the common cold. From dogs, we caught measles. Pigs and ducks passed on their influenzas, while smallpox and tuberculosis were transmitted to us from cattle. Today, we share around sixty diseases with dogs alone, and only slightly fewer with pigs, goats, sheep, horses, and cattle.

Furthermore, this process is still ongoing, as diseases continue to transfer from animals to humans. HIV originated with primates, and only transferred to humans around the start of the twentieth century. In 1994, a new virus called Hendra was discovered, after an outbreak that caused the death of fourteen horses. In the second outbreak later that year, two more horses died, along with their owner. Even more recently, both avian flu and West Nile virus have crossed over into human populations.

what was chimney
sweep's scrotum?

The occupation of a chimney sweep in Victorian London was a dirty and dangerous business. Child labor was frequently employed, as children were ideal for scurrying up and down narrow chimneys. The traditional image of a chimney sweep is a young boy whose face and clothes are caked in soot, but in fact sweeps usually worked naked, as clothes would have been a hindrance, liable to get snagged or damaged on the inside of the chimney. The first law attempting to improve the working conditions for chimney sweeps was passed in 1788, and it required that each chimney sweep could have no more than six apprentices, who had to be at least eight years old. It's astonishing to think that we're just a few generations away from a society that considered it progress to only allow children of eight and above to undertake dangerous physical labor.

Unsurprisingly, spending your days trapped in narrow, soot-filled chimneys carries a range of health risks. One such condition was known as "soot warts," which were blackened sores that would first appear on the scrotum, and then spread. There was no effective treatment, and victims would often have their scrotum removed, which led to infection and death. Soot warts were thought to be some kind of sexually transmitted disease, perhaps related to syphilis. However, in 1775, Percivall Pott established that the condition, which was unusually common among chimney sweeps, was a type of cancer, caused by exposure to carcinogenic soot.

This was an important discovery, not just for chimney sweeps, but for health care in general, as this was the first time that cancer had been demonstrably linked to an external factor. Before Pott's discovery, cancer was generally thought to be a systemic disease, caused by an excess of black bile. Thanks in part to Pott's work, there were increasing efforts to regulate chimney sweeping and improve working conditions, although children continued to be used for another hundred years.

Surprisingly perhaps, there are still many active chimney sweeps working today, although they rarely climb inside the chimney these days, and nakedness is almost certainly a thing of the past. Nowadays, chimney sweeps are more likely to describe themselves as "chimney technicians," whose services extend beyond cleaning and unblocking chimneys, to building and repointing chimney pots, repairing fireplaces, and fitting birdcages. There is also a long-held tradition that having a chimney sweep shake the bride's hand or blow her a kiss on her wedding day will bring her luck, so many chimney sweeps hire themselves out for just this purpose!

did the great fire of london wipe out the plague?

The conventional view once was that it was the Great Fire of London in 1666 that finally wiped out the plague. Epidemics of plague had recurred in London about once every generation since the mid-fourteenth century, but after the Great Fire there were no more significant outbreaks. The plague is believed to have been caused by the bacterium *Yersinia pestis*, which is transmitted via fleas and rats. Thus, the theory goes, when the Great Fire wiped out London's population of fleas and rats, destroying the dirty, infested wooden buildings where they bred, it therefore wiped out the plague, which never returned to London.

There is now some doubt about this theory, for a number of reasons. First, plague seems to have disappeared from most European cities at around this time, cities that had no equivalent fires. It now seems more likely the plague was already somehow losing its virulence at this point in time; that is, after a number of outbreaks, survivors and their offspring had become increasingly immune. It has been noted that survival rates tended to be progressively higher during each fresh bout of plague.

A further problem with the theory is that the Great Fire only affected the city of London, which was a small, enclosed area at the center, and not the

poorer suburbs. Only about a sixth of London's population lived in the city; the rest lived in the suburbs.
Furthermore, the suburbs were also the areas that
tended to be most affected by plague. In the Great
Plague of 1665, the vast majority of deaths occurred
in the suburbs, and not the city. Since the plague
chiefly affected the suburbs, rather than the city,
but the fire was only in the city, and not the suburbs, it's hard to see how the fire in one place could
have wiped out the plague in another.

Another objection is that the fire was quite slow-
moving. It lasted for five days, and initially it was
well enough contained for the Lord Mayor to dis-
miss it and go back to bed, commenting, "a woman
might piss it out." His approach was clearly proved
to be wrong, and later rightly vilified, but it does give
an indication of the slow rate of the fire's growth.
For the first few days, people didn't even try to flee
the walled city. Instead, those that vacated their
homes simply took a few possessions and moved to
a safe area, fully expecting to be able to return soon
after. Samuel Pepys found time to bury a cheese in
his garden for safekeeping. Presumably, a fair proportion of the rat and flea populations would have
been equally capable of escaping a slow-moving
fire, rather than necessarily dying in it.

The other enduring myth about the Great Fire
of London is that it only killed a handful of people,
but this too seems like it may be dubious. While
it is true that only five or six people are recorded

as dying in the fire; the deaths of the lower classes tended not to be recorded at this time. There were also accounts of foreigners being lynched by angry, rampaging mobs, as the fire was thought to be a Catholic plot, but none of these deaths, if there were any, are recorded. The firestorm reached extraordinary temperatures, thanks to the narrow streets, overhanging jetties, wooden buildings, and stores of gunpowder, fuel, tar, and other combustible materials. It became hot enough to melt the piles of imported steel lying along the wharves. It was hot enough to melt the iron chains and locks on the city gates. If the fire was hot enough to melt steel and iron, it was also hot enough to completely vaporize the bodies of anyone left behind or trapped, leaving no skeleton or remains.

The fire is also likely to have caused many deaths by slightly less direct means than people getting trapped and burned alive in the inferno. Many people were made homeless by the fire, and had to live in unsanitary and dangerous refugee camps. Many would have suffered from smoke inhalation, hunger, hypothermia, and burns. It's hard to imagine that significant numbers of the old, infirm, and children would not have died of some of these causes in the weeks and months following the fire. In one of the key eyewitness accounts of the fire, John Evelyn describes the "stench that came from some poor creatures' bodies." It's not certain that this refers to dead bodies, but it seems likely. One estimate suggests that the fire may have caused hundreds of unrecorded deaths, and perhaps even thousands.

why do we keep vials of smallpox?

The eradication of smallpox is regarded as one of medicine's greatest triumphs. Smallpox had blighted humanity since ancient times, consistently killing around 30 to 35 percent of those who became infected, with no treatment ever successfully developed. In the twentieth century alone, it is estimated to have caused somewhere between 300 million and 500 million deaths. Even as late as 1967, around 15 million people contracted the disease, with 2 million deaths. The last epidemic of smallpox took place in Bangladesh in 1975, but was quickly contained. In 1978, the last two recorded cases occurred at the University of Birmingham in England, as a result of a laboratory error. In 1979, the World Health Organization announced that smallpox had been completely eradicated.

But "eradicated" may not mean quite what you think it means. It doesn't mean that smallpox no longer exists; it just means that there are now no living carriers—smallpox only infects humans, so there is no possibility of there being any infected animals, either. There are still some confirmed vials of smallpox stored in government labs at the Centers for Disease Control and Prevention in Atlanta, Georgia, and the State Research Center of Virology and Biotechnology in Koltsovo, Novosibirsk Oblast, Russia.

Until fairly recently, there was a global consensus that these remaining vials should be destroyed, to rid the world of this potential blight. Bill Clinton signed off on a World Health Organization plan to destroy the last remaining stocks in 2002. Under George W. Bush, however, the United States changed its stance, due to fears about smallpox being used in bioterrorism. Anti-terror experts believe that a number of volatile nations may have got their hands on quantities of smallpox, possibly including China, Iran, Israel, North Korea, Serbia, and Pakistan. There are also reports that, in the relatively recent past, Russia secretly manufactured twenty tons of live smallpox virus, to be used as a biological weapon, and it's not clear what has happened to it.

America may be more vulnerable to smallpox today than it has been for perhaps 150 years. Current stocks of smallpox vaccine are believed to be low, and they may well be deteriorating in quality, or even ineffective. America stopped routinely vaccinating people in 1972, so few people born after that year will be protected, and the immunity of many older people may also have faded, as it's not clear how long smallpox immunization lasts.

Thus, rather than destroying the remaining vials, the U.S. plan now seems to be focused on retaining the samples while developing a more effective vaccine, as well as antiviral drugs to actually treat the disease.

what is the "disease of kings"?

 The answer is gout, which has traditionally been regarded as an affliction of the wealthy and privileged, due to its association with rich foods and alcohol. Gout is an unpleasant, painful condition caused by an excess of uric acid in the blood. This crystallizes and forms deposits in joints, tendons, and surrounding tissues, causing painful, tender red swellings, most often at the base of the big toe. Because of gout's association with the wealthy and well-fed, newspaper cartoons of the Industrial Age were often populated with portly, red-faced gentlemen, hobbling with painful, bandaged feet. In particular, gout has long been thought to result from regular consumption of port. The gentleman's clubs of London's Pall Mall would even provide a small velveted and tasseled footstool, so that the painfully tender toe of a gout sufferer could be lifted, and thus somewhat relieved.

Surprisingly, a recent study came to the conclusion

that in fact drinking port could protect people from gout, as the condition was far more likely to be caused by drinking beer. Moreover, gout affects not only the most eminent humans. Scientists have recently discovered that *Tyrannosaurus rex*, the king of the dinosaurs, also suffered from gout.

chapter three

dodgy diagnoses

Optimistic lies have such immense therapeutic value that a doctor who cannot tell them convincingly has mistaken his profession.

GEORGE BERNARD SHAW

what was "hepatomancy"?

Hepatomancy, which is also known as hepatoscopy, was an ancient medical system devised in Mesopotamia, in modern-day Iraq. The way it worked was that patients would be diagnosed by a physician, who would make his judgments by inspecting not the patient, but the livers of some sacrificed animals. The liver was believed at this time to be the source of our blood, which meant it was the source of life itself. Following this logic, people therefore believed that the will of the gods could be divined by inspecting the livers of sacrificed sheep. (I know, me neither.) Each section of the liver was believed to correspond to a particular deity, and clay models of sheep's livers, which were presumably used by doctors to help them to interpret the signs, have been found dating back as far as 2050 B.C.

of the four "humors," what was particularly special about "melancholy"?

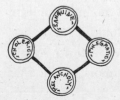

The Hippocratic corpus held that human health depended on the body's four types of fluid, which were called the four "humors." It was believed that every aspect of our health could be explained by the balance of these humors, which were blood, phlegm, yellow bile, and black bile, also known as "melancholy." Each humor was believed to play a crucial and distinct role in the body's health. Blood was the source of strength and vitality. Bile was the gastric juice, which was needed for digestion. Phlegm was seen as a lubricant and coolant. And black bile was responsible for darkening the other fluids, such as when blood or stools become darker.

These four humors were at the center of a complex network of signs and theories, which were believed to explain every aspect of a person's health and personality. Each humor was associated with one of the four elements that the Greeks believed to comprise the universe. Blood was hot and agitated, like air. Yellow bile was hot and dry, and so connected to fire. Phlegm was linked to water, and black bile was cold and dry, like earth.

These types of analogies were extended to link the humors to astrology, the seasons, and different character types. Phlegm, for example, was cold and wet, which meant it was associated with winter, when people catch colds and chills. People with excess blood were regarded as lively, confident, and robust, or "sanguine," while those with too much phlegm might be lazy and easygoing, or "phlegmatic."

So what was special about melancholy? Simply, it's the only one of the four humors that doesn't actually exist. Our bodies do contain blood, phlegm, and yellow bile, and these were easily observed by doctors then and now. Yet with no understanding of why blood, urine, or feces might get darker at times, or why some people have darker complexions than others, the Greeks assumed that some as yet unknown fluid must be responsible, and so credited these changes to an unseen substance called black bile, or melancholy.

what is "laudable pus"?

Before bacterial infection was understood, almost all surgical wounds would become infected. The main distinction between wounds was between those that produced a creamy, yellow pus, or suppuration, and those with a thin, watery discharge. Since the time of Galen, doctors had believed that this suppuration was indispensable for healing, as pus derived from poisoned blood, which needed to be expelled from the body. This view held sway until the late nineteenth century, because before germs were understood, it seemed to make sense. Although wounds that produced yellow pus tended to take months to heal, physicians noticed that those patients had a higher survival rate than those whose wounds were pus-free. Naturally enough, doctors concluded that yellow pus was good for you.

As a result, doctors would actively seek to encourage infections after surgery or injury. One early medical guide advised doctors to "get wool as greasy as can be procured, dip it in very little water, add one-third wine, boil to good consistency," and then insert it into the wound. Even in the late nineteenth century, it was widely accepted that cancer patients would benefit from an infection to heal the wound, after cancer surgery had taken place.

Some doctors went even further, recommending gangrene. The Parisian Stanislas Tanchou argued, "It is remarkable that . . . gangrene has caused the largest number of cures. Gangrene may be considered as a therapeutic agent, whether it occurs spontaneously or is induced medically." In truth, the only thing that gangrene did regularly cause was the amputation of people's arms and legs.

why are doctors in old paintings often holding up a flask?

The answer is that throughout much of early history, diagnosis was largely conducted by inspecting the patient's urine. The practice is known to have been conducted as far back as ancient Egypt, Babylon, and India. Uroscopy, to give it its proper name, was not a central part of Hippocratic medicine, which relied more on observations made by inspecting and touching the patient, but it somehow became widely prevalent in medieval Europe. If a doctor is present in a medieval painting, he will often be holding a large, bulbous flask up to the light, as this flask was as much the symbol of his profession then as a stethoscope and a white coat are today.

Urine diagnosis was thought to be so powerful that patients were not even required to visit their doctor themselves; often they would simply send a urine sample from their sickbed, to be diagnosed in their absence. Doctors would do this, if possible, by inspecting the urine in various states, first when it was hot and fresh, and then later when it had cooled down, and had presumably taken on a different consistency.

Doctors used detailed color charts against which they would match and compare the urine to diagnose a wide range of conditions, including kidney disease, jaundice, and urinary tract infection. Then, on the basis of this, along with the patient's astrological chart, they would determine which humoral imbalance needed to be rectified. Thick urine could mean the patient was suffering from dropsy, colic, or an excess of phlegm; dark urine indicated an excess of black bile, and so on. But not all conditions could be judged by sight alone. The smell was also instructive, as was the taste. Diabetes, for example, makes the patient's urine sweet, so doctors would routinely taste their patient's urine. They really did earn their money in those days.

Today, of course, we know that only limited

information can be learned from a person's urine, and it is not the central plank of diagnosis. Doctors do still analyze urine, testing its concentration and pH, as well as its levels of drugs, blood, sugars, fats, bilirubin, and hormones. And in fact some of these tests can help to diagnose some of the very conditions that the medieval doctors were looking for, including kidney stones, urinary tract infections, and diabetes. Chemical analysis of urine is even undergoing something of a resurgence as a field of study, under the rather fancier name "metabolomics." However, tasting patients' urine has fallen somewhat out of fashion among doctors.

what was a "plague doctor"?

During the fourteenth century, the Black Death swept through Europe, killing 30 to 50 percent of the continent's population, in one of the most devastating pandemics in history. There were three forms of plague: pneumonic, characterized by fever and bloody phlegm; septicemic, which caused fever and patches of purple skin; and the most common form, bubonic, which made grotesque, pus-filled buboes appear around the groin, neck, and armpits. All three forms were highly fatal, and highly contagious. The plague returned about once every generation until the seventeenth century.

When the plague hit, the wealthy would flee London, believing that the disease was caused by miasmas of bad air. Knowing that they could do nothing for plague victims, most doctors also fled the city, and so in their place "plague doctors" were appointed. These were not really doctors at all; their duties generally consisted of visiting the homes of the sick, to assess whether or not they were afflicted with plague. If there were plague victims inside, the house would have to be quarantined, with no one allowed in or out, and a red cross marked on the door. Although they were not qualified, plague doctors were often very well paid, to compensate them for the considerable risk they faced of becoming infected themselves.

To limit this risk as far as possible, plague doctors wore a protective suit that has become an iconic, sinister image of plague. The suit comprised a long, black, high-necked overcoat, which extended to the feet, and leather breeches, similar to a fisherman's waders, to protect the legs and groin from infection. A wide-brimmed hat was worn, as this was a symbol of doctors. Most distinctively of all, plague doctors wore red glass eyepieces, and a kind of long-nosed gas mask, in the shape of a bird's beak. The red eyepieces were meant to ward off evil, while the beak was designed to protect the wearer from the "bad air" that was thought to cause plague.

The end of the beak was often filled with herbs

or spices, in the hope that these might overcome the miasmas. The whole suit would be covered with suet or wax, as a further protection. Finally, plague doctors would carry a wooden cane, which may have been used to gesture to the patient's family (as communication must be fairly restricted in a wax-covered gas mask with a beak full of herbs), or to physically move the patient's body.

why did claude bernard's wife leave him?

Claude Bernard was a leading French scientist who is regarded as a pioneer of physiology, and a key figure in the history of medicine. He discovered the role of the pancreas in digestion, and production of glycogen by the liver. He succeeded in bringing the rigor and experimental approach of science to medicine, which at the time was still largely based on superstition, anecdote, and received wisdom. When he died in 1878, Bernard was given a public funeral, becoming the first French scientist to be granted such an honor.

Throughout his career, Bernard argued for the importance of experimentation and observation. This was controversial, particularly when it came to

Bernard's approach to vivisection, which is experimentation on live animals (its name comes from the Latin words for "live cutting"). Many of Bernard's contemporaries were horrified by the cruelty involved in experimenting on live animals, which was often conducted without anesthetic. Bernard recognized that animal testing was deeply unpleasant, but he felt it offered lifesaving results which were too important to ignore. He wrote, "The science of life is a superb and dazzlingly lighted hall which may be reached only by passing through a long and ghastly kitchen." Others, however, felt that such horrors could never be justified.

In 1845, Bernard married Françoise Marie Martin, who was known as Fanny. It was a marriage of convenience, as Fanny came from a wealthy family, and so the match allowed Bernard to pursue his scientific career. As time progressed, though, Fanny became increasingly disgusted by Bernard's practice of vivisection, as did the couple's daughter. In 1869, the couple were officially separated, with Fanny citing his cruelty to animals as being a major reason for the breakup of the marriage. Fanny went on to become a vocal campaigner against the use of vivisection.

why was queen victoria's hernia never diagnosed?

In 1881, Sir James Reid was appointed as the personal physician to Queen Victoria, who was then a reasonably fit, if somewhat overweight sixty-two-year-old. Victoria was something of a hypochondriac, so the pair had a close, involved relationship. The queen would summon Reid six times a day, call him back from holidays, and keep him intimately briefed on every detail of her constitution. While he was on his honeymoon, she wrote to brief him: "The bowels are acting fully." Even after he married, Reid was not allowed to live with his wife, as Victoria needed him to be at her beck and call. When she died, Reid was given a task of the highest secrecy and delicacy: he was to put a lock of hair from Victoria's close companion John Brown into her hand in the coffin, along with a photograph of Brown. The point of all this detail is to emphasize that Victoria and James Reid clearly had an extremely close, trusting relationship.

After Victoria's death in 1901, Reid inspected her body, and found that she had a hernia and a badly prolapsed uterus, which had never been diagnosed. The reason they had never been diagnosed was that Reid had never been allowed to see Victoria with her clothes off, as propriety forbade it. He had never even been able to place a stethoscope against her chest, and he never once saw her in her own bed until six days before her death.

This kind of primness sounds like the kind of thing that could only come about through the absurdities of royal protocol, but in fact it was fairly typical of how normal doctors of the period operated, and had done for centuries. Galen had recommended taking the patient's pulse at the wrist, as this did not necessitate any undressing. Through the Middle Ages, doctors seemed to gradually retreat further from physical contact with their patients. The eighteenth-century German doctor Johannes Storch, who was a leading authority and widely published, would rarely even meet his patients. Instead, he carried out most of his diagnosis and treatment through correspondence and intermediaries. As late as the 1890s, some family doctors in the United States would do nothing more tactile than taking the patient's pulse and inspecting his or her tongue.

what's the difference between a moron and an imbecile?

The short answer is: about 25 IQ points, or at least that used to be the answer. Today, the words *moron* and *imbecile* mean the same thing: They are both pejorative, insulting terms that simply convey the concept of "person of low intelligence." But this was not always the case. In the first American IQ tests, "moron," "imbecile," and "idiot" were neutral, inoffensive terms that each had a specific and distinct meaning.

The very first IQ test was designed for a specific, practical purpose. It was developed in 1905 by two French psychologists, Alfred Binet and Theodore Simon, and thus became known as the Binet-Simon scale. The system comprised a series of thirty tasks of increasing complexity. The simplest tasks included asking the child to shake the examiner's hand and follow a lit match with his or her eyes. The most difficult tests included remembering a series of seven random numbers, and finding three rhymes for the word *obeisance*, which is presumably easier if you are actually French, as these children luckily were. The tests were based on observations of an average child's level of abilities at each age, and thus created the concept of a "mental age," distinct from the child's actual age.

As I stated, the purpose of this system was a purely practical one: to provide a fair and consistent way of measuring which schoolchildren had special learning needs, so that these children could be taught separately. The invention of the Binet-Simon scale was therefore designed to address the problem of teachers choosing to exclude children who were troublesome, but not developmentally challenged.

Binet was careful to recognize the limitations of his system, as he believed that intelligence was not something inherent and fixed, but much more fluid and complex. He wrote:

> I do not believe that one may measure one of the intellectual aptitudes in the sense that one measures length or a capacity. Thus, when a person studied can retain seven figures after a single audition, one can class him, from the point of his memory for figures, after the individual who retains eight figures under the same conditions, and before those who retain six. It is a classification, not a measurement. We do not measure, we classify.

When the system crossed the Atlantic, it was taken up by groups with very different goals. The first translation of Binet's system was carried out by Henry H. Goddard, who was a leading American psychologist. He was also a eugenicist, which meant he believed society should be run by the most intelligent people, and that those of the lowest intelligence should be institutionalized or sterilized. Goddard developed IQ tests, based on Binet's scale, which categorized adult participants as gifted (IQ of 130+), normal (71–129), moron (51–70), imbecile (26–50), or idiot (0–25).

While there are many things we might criticize Goddard for, his use of language shouldn't be one.

At the time, *moron* and *imbecile* were not insulting terms. In fact, the word *moron* didn't even exist. Goddard invented it in 1912, from the Greek word *moros*, which means "dull." It was only because these tests became common and widely known that these words took on a pejorative meaning in the wider language.

This shift in meaning illustrates an interesting concept, the "euphemism treadmill," in which neutral words that come to refer to things perceived as negative gradually acquire a derogatory, taboo meaning. Initially, *moron* was not an insulting term, but by the 1960s its popular meaning had changed, and so psychologists began using new, inoffensive terminology, describing these lower categories with the more enlightened classifications of mild, moderate, severe, and profound retardation.

However, as you've no doubt already noted, in time those terms also became unsuitable, as did the next politically correct idiom: "mentally handicapped." Today, the polite form is to refer to a person as having "special needs" or "learning difficulties," a construction in line with our contemporary desire to avoid defining people purely by their condition. It seems that no matter how carefully we try to design neutral, inoffensive language for sensitive matters, the meaning of such language is always likely to shift and become pejorative in time.

what is a lancet?

A lancet is a medical device that was traditionally used for bloodletting. From the time of Hippocrates until the nineteenth century, illness was held to be caused by an imbalance of humors, and the main tools at a doctor's disposal remained roughly unchanged. They were purging (which means in-duced vomiting), sweating, enemas, and perhaps most often of all, bloodletting. Bloodletting was so central to medicine for so much of its history that the name given to the leading medical journal was *The Lancet*, and it still carries this name today.

There were various types of lancet, including spring loaded lancets, thumb lancets, fleams, and unpleasant-sounding scarificators, but they all essen-tially consisted of one or a number of small blades or spikes used to cut into a vein or artery. As well as lan-cets, there were various other devices used for blood-letting. Some doctors used "fire cupping," in which a small bowl would be heated, and then held against the skin at the site of incision, creating a vacuum that would draw out the blood. Many doctors used leeches, to such an enormous extent that France had to import millions of the creatures, having harvested its own population to the brink of extinction.

Bloodletting was recommended by doctors for almost any ailment. Blood was believed to originate in the liver, and then grow stagnant as it built up in the body. Consequently, blood loss was generally seen as a good thing, and nosebleeds, varicose veins, and hemorrhoids were all to be encouraged, because they helped to get rid of excess blood. Patients were often bled until they fainted; if the patient hadn't fainted, that was usually a sign that you hadn't taken enough blood.

Some of the uses of bloodletting were quite absurd. Galen's recommended cure for a patient suffering from blood loss was, unbelievably, bloodletting. Before surgery, patients would routinely have blood removed, as would mothers before childbirth. Before an amputation, doctors would estimate the amount of blood that the limb contained, and then remove this amount, before then removing the limb (which was, of course, still full of blood).

In the vast majority of cases, bloodletting did considerably more harm than good, and many patients died as a result. What's astonishing is that bloodletting continued to be used and advocated, even as the science of medicine advanced, and the theory of the humors increasingly became redundant. In 1628, William Harvey demonstrated that blood circulated around the body, effectively disproving the logic of bloodletting, yet the practice continued. In the 1830s, bloodletting was shown to be useless for pneumonia and fevers, for which it was commonly used, but nonetheless the practice continued. In 1857, John Hughes Bennett produced detailed statistical evidence comparing bloodletting with both nontreatment and placebo treatment, and clearly demonstrated that nontreatment was better than bloodletting. Even then, bloodletting endured.

With each advance in the science of medicine, bloodletting became increasingly anachronistic, and yet it persisted, because although new knowledge was being rapidly developed, there were hardly any new cures or treatments, and doctors had to continue to offer treatments of some kind. One medical historian has estimated that, despite all the advances of the eighteenth and nineteenth centuries, there was no meaningful progress in the effectiveness of treatments until Joseph Lister's discovery of antisepsis in 1867. This may explain why, as late as 1923, a leading medical textbook was still recommending bloodletting.

Although it was mostly harmful, bloodletting was beneficial for a handful of conditions. Edema, which used to be known as dropsy, is a type of fluid retention that can cause severe heart and lung problems. Today, edema is treated using diuretics and vasodilators, but in the past bloodletting did have a useful effect. Polycythemia is a disease in which your body produces too many red blood cells, and so bloodletting is still used for this. Other conditions that still call for bloodletting include the enzyme disorder porphyria, and hemochromatosis, in which your body absorbs too much iron.

does chocolate give you pimples?

 There is an enduring myth that chocolate causes zits, but the evidence does not support it. A recent meta-analysis, which combined the results of twenty-one separate observational studies and six clinical trials, found no link between chocolate consumption and acne. There was also no link between acne and greasy foods such as burgers, fries, and pizza (although of course none of this undermines the more general importance of a healthy, balanced diet). The study did, though, find a possible link between acne and cow's milk, as well as foods with a high glycemic index, such as highly refined bread and cereals. Chocolate has a low glycemic index.

Overall, the evidence continues to suggest that the main cause of acne is not diet, but simply hormones. During adolescence, our bodies produce more of the hormone androgen, and one of its many effects is to stimulate our sebaceous glands to produce more sebum and grow in size. Pimples occur when our hair follicles get blocked with dead skin, causing sebum to build up behind the plug. Ninety-six percent of teenagers are affected by acne at some point, and in many sufferers it continues into adulthood. The other important factor is simply genetics. Acne will tend to run in families, as you are likely to have a similar skin type to your parents.

did president harrison die of a cold because he refused to wear a hat during his inauguration?

William Henry Harrison was a notable U.S. president for a number of reasons. He was elected in 1841, but died just 32 days later, making him the first president to die in office. His tenure remains the shortest on record, significantly outdoing second-place man James A. Garfield, who was shot by an assassin after a full 199 days in office. Harrison is also notable for being elected with the highest share of the eligible vote of any president, although this is partly due to the high voter turnout at that particular election.

At the time, Harrison was also the oldest man to become president, at sixty-eight years old, and his detractors mocked him for this, referring to him as "Granny." Perhaps to emphasize his health and vigor—he was, after all, a war hero—Harrison delivered the longest inauguration speech on record, at one hour and 45 minutes, on March 4, 1841. It was a bitterly cold day, it may even have been snowing, but despite the weather, Harrison refused to wear a hat or overcoat, and the entire address was conducted outside. As a result, Harrison caught a cold. A month later, he died of pneumonia.

Yet these events may not be quite as connected as the sneaky phrasing of that last paragraph might suggest. Harrison fell ill with his cold on March 26, more than three weeks after his inauguration, which suggests that it couldn't have been caught on that day. Cold symptoms usually begin within two to five days of infection.

The cold also couldn't have been caused by the weather, because, despite what many people believe, colds are not caused by low temperatures, but by a virus. It's true that we get more colds in the winter, but that's not because we're outside in the cold. On the contrary, we catch more colds in winter because we're huddled up indoors, with the heating on and the windows shut. In the winter, we find ourselves in closer proximity to other people, who might infect us with colds, and our warm, poorly ventilated homes are ideal for viruses to thrive in. This won't be of much comfort to President Harrison, but at least his death wasn't his own fault for not wearing a hat.

chapter four

curious cures

He's the best physician that knows the
worthlessness of most medicines.

BENJAMIN FRANKLIN

why do boxers drink
their own urine?

 In 2009, boxer Juan Man-
uel Márquez allowed HBO's
cameras to follow him as
he trained for his fight with
pound-for-pound champion
Floyd Mayweather, Jr. At one
point in the show, viewers
saw Márquez drink a glass of his own urine, which
he explained was a regular part of his training
regime. According to the boxer, urine contains lots
of vitamins, which are expelled by the body, and so
he drinks it for his health.

In fact, drinking urine, or "urine therapy," to
give it its proper name, is surprisingly common
among fighters. The Brazilian mixed martial arts
(MMA) light heavyweight Lyoto Machida drinks his
own urine every morning, as do his whole family.
Fellow MMA fighter Luke Cummo also advocates
urine therapy, claiming it gives him the edge over
his opponents, although since his record so far is
six wins and six defeats, he may not be the best ad
for the practice.

After winning his WBC World Heavyweight title
fight against Nigerian Samuel Peter in 2008, Vitali
Klitschko revealed his own special use for urine. He
wraps his baby's wet nappies around his fists after

a fight, to prevent swelling. He explained, "Baby wee is good because it's pure, doesn't contain toxins, and doesn't smell." At the time of writing, Klitschko is still the WBC champion, so perhaps it does help!

In fact, urine drinking is not just practiced by punchdrunk pugilists. Many devotees of alternative medicine swear by the practice, which has gone on for thousands of years. Indian holy men would drink their own water, as well as massaging it into their skin. The Koran bizarrely recommends the drinking of camel urine for its medicinal properties. In China, the urine of young boys is drunk by adults, as it is believed to have health-giving properties, and babies' faces are washed in urine, apparently to protect their skin. As recently as 1978, the prime minister of India was publicly advocating urine therapy as the ideal solution for those millions of poor Indians who could not afford to see a doctor.

So, is there anything to it? Sadly (or not, I suppose, depending on how keen you are to drink urine), the answer seems to be no. The practice is fairly harmless, as urine is generally sterile, and any infection that might be carried in the urine would already be present in the body, assuming a person only drank his or her own urine, rather than somebody else's. Márquez is right to say that urine may contain vitamins, but these are vitamins that the body has already expelled, because it has no use for them. The only meaningful effect is likely to be a bowel movement, or even diarrhea, as the urea in urine has a laxative effect. Overall, then, there is no evidence that any form of urine therapy is beneficial. And as for the fight, Márquez lost to a unanimous points decision, 120–107, 119–108, 118–109, which means that Mayweather won almost every round.

ah, but what about jellyfish stings? everyone knows urine is good for them . . .

Well, we all remember that episode of *Friends*, when Joey and Chandler have to pee on Monica's leg to ease the pain of a jellyfish sting. However, although this supposed remedy is well known, in fact the evidence seems to suggest that urine wouldn't have done any good at all, and might even have made Monica's pain worse. A jellyfish's tentacles are covered with tiny stinging cells called nematocysts. These are fired into the jellyfish's prey at enormous speed, injecting a powerful venom.

When a person has been stung by a jellyfish, there are two main priorities. First, any remaining parts of the tentacle have to be removed, without triggering the nematocysts to release more venom. The nematocysts will release venom in response to freshwater, certain chemicals, and physical pressure, so one recommendation is to use a towel or sand to carefully remove any remaining parts of the tentacle. Touching the tentacle with your bare hands is not a good idea, as it will only result in more stings.

The second job is to wash away the remaining nematocysts, but this is a difficult task. Freshwater will make them release more venom, as they react to a change in the concentration of salts. Various other liquids have been recommended, including Coke, red wine, and ammonia, but there is very little evidence to support any of them. As you might imagine, people are not exactly queuing up to take part in scientific tests involving getting stung by a jellyfish (although, amazingly, some volunteers have been found!). Based on what evidence there is, it seems that weak acetic acid, also known as vinegar, is perhaps the best choice for washing the sting, as this acid is known to neutralize the nematocysts of many types of jellyfish, including the deadly box jellyfish. Some other types of jellyfish react badly to vinegar, and the sting becomes more painful, so seawater may be preferable in these cases.

The theory that urine helps to ease the pain of a jellyfish sting may have developed because urine is known to be mildly acidic. But the acidity of urine is very weak, and it's likely therefore that the nematocysts would react to urine as if it were freshwater; in other words, it would make them release more venom. Therefore, washing a jellyfish sting with urine is not recommended.

Once the skin has been rinsed, either with vinegar or seawater, the remaining nematocysts can be removed by shaving the area, using a razor or a knife. Most jellyfish stings are painful, but not fatal, and the pain usually disappears after about twenty-four hours. Having your best friends pee on you, on the other hand, is likely to sting for much longer.

why was urine used to wash battlefield wounds?

Urine is also believed to have been used for centuries as an antiseptic, to clean battlefield wounds, when no clean water or other antiseptics were available. It may sound unpleasant, but urine is usually sterile, unless the donor has a urinary tract infection, which means it was probably a reasonably effective solution, so to speak.

Urine was infinitely preferable to some of the alternative battlefield balms. Before the sixteenth century, it was routine to pour boiling oil onto serious battlefield injuries, to cauterize the wounds and stop the bleeding. One day, the great French surgeon Ambroise Paré was tending the wounded on a battlefield, when he ran out of boiling oil. He remembered an old ointment he had read about, which had supposedly been used in Islamic medicine, and decided to give it a try. He treated the remaining patients with this ointment, and then left for the night.

When he returned the next day, Paré had inadvertently performed a scientific test. The soldiers who had been treated with boiling oil were still in agony, and showing no improvements. Those who had received the soothing ointment were much improved, thanks largely to the antiseptic properties of turpentine.

was hitler addicted to crystal meth?

Bizarrely, it seems that he was. Adolf Hitler was a hypochondriac Every day, he would have his doctor Theodor Morell give him vitamin injections into his buttocks. As World War II progressed, Hitler was having as many as five injections a day, and Morell was lacing them with methamphetamine, which is also known as crystal meth, to keep the Führer alert and energetic. Thanks to these injections, Hitler was "fresh, alert, active, and immediately ready for the day . . . cheerful, talkative, physically active, and tending to stay awake long hours into the night." Albert Speer felt that Hitler's methamphetamine addiction was one reason for his rigid tactics in the later stages of the war, when he would refuse to allow troops to make any tactical retreats, regardless of the circumstances.

Although Hitler's drug use may sound surprising, in fact amphetamines were widely used during World War II. German soldiers were fueled by an amphetamine called Pervitin, with the army supplying millions of tablets to the troops, particularly during the blitzkrieg attacks on Poland and France. The British and Japanese armies also distributed huge volumes of amphetamines to their soldiers. The U.S. armed forces didn't authorize amphetamine use on a regular basis until the Korean War, but many troops stationed in Britain became familiar with Benzedrine—one estimate suggests that 80 million Benzedrine pills were supplied to U.S. servicemen by the British, with the same amount again being provided by U.S. medics.

Although Hitler's drug addiction is well documented, there are many other theories about his health that are more speculative. Various historians have argued that Hitler suffered from syphilis, Parkinson's disease, Asperger syndrome, irritable bowel syndrome, irregular heartbeat, skin lesions, and a range of mental health disorders including borderline personality disorder and schizophrenia. One enduring rumor held that Hitler had only one testicle, perhaps as a result of an injury he suffered during World War I, when he was shot in the groin.

After Hitler's death, a Russian autopsy confirmed that he was missing his left testicle, which seemed to prove that the popular wartime song "Hitler Has Only Got One Ball" was, amazingly, correct!

But there are strong suspicions that the Russian autopsy was a work of propaganda—for one thing, Hitler's body is believed to have been cremated in the bunker where he shot himself, which would presumably not leave much of a scrotum to be examined. Furthermore, there is evidence that the early popular versions of the "one ball" song laid the accusation at the groin of Hermann Göring, with the song supposedly beginning, "Göring has only got one ball, Hitler's are so very small . . ." Although the order of the protagonists was later reversed, it would seem to be a bizarre coincidence for a song to appear, making such a bizarre and specific claim about Göring, which then turned out to be true of Hitler instead. Opinion is divided therefore—although that's not to suggest that historians are losing much sleep over the issue—but my suspicion is that it's a myth.

how did agatha christie's *the pale horse* save lives?

On its release in 1961, Agatha Christie's murder mystery *The Pale Horse* marked something of a departure from her usual, classic formula. The book didn't feature either of Christie's two celebrated detectives, Hercule Poirot and Miss Marple, and it was set in a distinctly contemporary, modern setting, rather than the standard country manor house which had become such a cliché of the genre. It even features a fistfight in a pub.

Like many of Christie's novels, the book features murder by poisoning. Christie herself had worked as a nurse and a pharmacist in her youth, which may have sparked her enduring fascination with poisons. In the case of *The Pale Horse*, the poison of choice was thallium, which is a fairly common metal. Thallium is a very effective poison, as many of its salts are odorless, tasteless, and easily dissolved. The effects of thallium poisoning take time, which is also useful for a would-be poisoner, and they are difficult to distinguish from other conditions such as epilepsy, encephalitis, and neuritis. Symptoms of thallium poisoning include lethargy, numbness, blackouts, slurred speech, and hair loss—in fact, in the 1930s thallium was sold over the counter as depilatory product, under the exotic-sounding name Koremlou Creme.

In 1975, fourteen years after the book's release, Christie received a letter from a woman in Latin America, thanking her for saving her friend's life. The writer explained that she had noticed the symptoms of thallium poisoning in a friend, thanks to having read *The Pale Horse*, and had consequently saved this unwitting friend, who was being slowly murdered by her husband.

In fact, *The Pale Horse* seems to have saved a number of lives since its publication. In 1977, a nurse at Hammersmith Hospital in London was treating a child from Qatar who was suffering from a mysterious illness. Luckily, the nurse had read *The Pale Horse*, and so spotted the symptoms of thallium poisoning, saving the child's life. In 1971, the notorious poisoner Graham Frederick Young was caught, thanks to a doctor connected with the case who recognized the symptoms of thallium poisoning, thanks again to *The Pale Horse*.

why did ancient egyptians apply moldy bread to wounds?

From the time of the ancient Egyptians onward, the principle that moldy bread could be used to keep cuts clean was a common and widespread tenet of folk medicine. Today, we know why it may have worked: It's because fungi can have antibacterial properties. This principle was discovered by Louis Pasteur, the father of bacteriology.

Scientists already knew that bacteria could be killed using chemicals, but in 1877 Pasteur discovered the possibility of using biological agents, when he observed that common molds slowed the growth of *Bacillus anthrasis*, the bacterium which causes anthrax. There is even some speculation that Pasteur identified this strain of mold as penicillin, many decades before penicillin is generally held to have been discovered in 1928.

what was the "powder of sympathy"?

Most of what was regarded as conventional medicine in the seventeenth century would strike us as wacky and bizarre today, but even in this context of superstitious quackery, Sir Kenelm Digby's "powder of sympathy" seems wonderfully insane. The powder was designed for a very specific purpose, to heal rapier wounds, and it contained a whole pharmacopoeia of bizarre ingredients, including earthworms, pigs' brains, iron oxide, and powdered mummy. This last ingredient was exactly what it sounds like: bits of mummified corpses, ground into a powder. Use of powdered mummy in medicine was actually fairly conventional at this time. Doctors had been prescribing powdered mummy for centuries for all kinds of ailments—a practice that seems to have started as a result of people in the twelfth century observing that mummified corpses were extremely well preserved.

As if this story weren't batty enough, this salve was not even meant to be applied to the wound itself, but rather to the weapon which had caused the wound.

Digby believed that applying this strange concoction to the offending rapier would somehow encourage the wound itself to heal, through "sympathetic magic." Furthermore, Digby was not some obscure crank. On the contrary, he was a renowned courtier, a diplomat, and a leading intellectual. His book on the powder of sympathy went through twenty-nine editions. He was also the first person to observe that plants require oxygen (or "vital air," as he termed it), and he invented the modern wine bottle. Digby's correspondence with the legendary mathematician Fermat contains the only extant copy of any of Fermat's mathematical proofs—a demonstration that the area of a Pythagorean triangle cannot be a square, through Fermat's method of infinite descent.

Another use that Digby proposed for his powder of sympathy was that it should be used by sailors to determine their longitude at sea, which was an extremely important and, as yet, unsolved problem. Digby's suggestion was that a bandage should be taken from a wounded dog, which would then be taken to sea (the dog, not the bandage). At a set, pre-agreed time each day, the bandage (on land) would be placed into the powder of sympathy, and through sympathetic magic, this would cause the dog (at sea) to yelp. Thus, the crew on board the ship would effectively get a daily alarm call telling them the correct time, allowing them to calculate the ship's longitude. It is not known whether or not this method was ever tried. Of course, one of the downsides of this method would be that, on long journeys, the dog would have to be frequently rein-jured, to make sure the wound remained effective.

which disease was thought to be cured by leading the patient, wearing a donkey's halter, three times around a pigsty?

According to a book of Irish folk traditions, it was believed that an effective cure for mumps was to put a donkey's halter over the patient's head, and lead the patient three times around a pigsty. Why anyone believed this might actually work, however, remains a mystery.

what was vin mariani?

Vin Mariani was a health tonic, created in 1863 by a chemist, named Angelo Mariani. He was inspired by an article detailing the powerful effects of cocaine, which had recently been synthesized from the coca plant, and so he decided to produce a wine that would harness these effects, by treating regular Bordeaux red wine with coca leaves. Mariani's tonic contained .0002 ounces (6 mg) of cocaine per fluid ounce (28 ml) of wine, although he later upped the rate to compete with stronger rivals.

The tonic was an enormous success, perhaps unsurprisingly. Mariani's advertising played up the tonic's health-giving properties, claiming the drink was endorsed by eight thousand doctors, and ideal for "overworked men, delicate women, and sickly children." It was enjoyed by many prominent people, including Thomas Edison, Queen Victoria, the Czar and Czarina of Russia, Pope Saint Pius X, and Pope Leo XIII, who even appeared in an advertisement for the wine, and awarded it a Vatican gold medal, whatever that is.

Interestingly, Vin Mariani's appeal was not just down to it containing alcohol and cocaine, two pleasurable drugs in their own right. It also benefited from the way in which they react with each other. When administered with alcohol, cocaine produces a potent psychoactive metabolite called cocaethylene, which is only produced by the two drugs taken together, not by cocaine on its own. Today, cocaethylene is becoming an increasing concern among health professionals, as there is growing evidence that heavy drinking combined with cocaine use is causing serious heart problems among the under-forties, because of the effect of cocaethylene.

Vin Mariani was successfully exported to the United States, where it sold extremely well, and inspired a certain John S. Pemberton to come up with a similar product of his own. In 1885, Pemberton began selling Pemberton's French Wine Coca, but prohibition laws were soon passed (although of course these only restricted the sale of alcohol, not cocaine). Pemberton responded by developing a non-alcoholic version of his drink, which he decided to call Coca-Cola, and this too went on to become reasonably popular, I believe.

what is a clyster?

A clyster is an archaic word for an enema, which is the introduction of water or some other liquid into the rectum and colon. Clysters are thought to have been performed since the time of ancient Egypt, supposedly inspired by the ibis bird, which uses its beak to flush salt water into its cloaca. Clysters became an extremely fashionable treatment between the seventeenth and nineteenth centuries among the wealthy. They were used to treat constipation, along with a wide variety of other ailments, although they were almost certainly useless for any complaint other than constipation.

A conventional enema would consist of warm water, possibly with added salt, baking soda, or soap. Some doctors would perform more elaborate enemas, adding coffee, bran, herbs, honey, or chamomile. To perform the enema, the doctor would insert a large syringe deep into the anus, beyond the sphincter muscles. The syringe would have a large bulb attached to the top, which would contain the enema liquid. In earlier times, a pig's bladder was often used for this purpose. The bulb was then squeezed, forcing the mixture through the syringe and into the colon.

In high society, enemas became enormously popular, with aristocratic hypochondriacs taking three or four delicately scented enemas a day. The patient would kneel, with their bum in the air, to give the doctor the best access. Some of the more refined women would try to maintain their privacy by kneeling behind a screen, which blocked the doctor's view of all but the anus, in a bizarre attempt to preserve their modesty.

Louis XIV of France was said to have had more than two thousand enemas over the course of his reign. He also set up a special team of detectives to investigate the spate of murders-by-enema-poisoning that were becoming a growing problem. On one occasion, the visiting Duchess of Burgundy shocked the King by taking an enema in his presence, as if it was the most natural thing in the world. It's little wonder that scenes featuring enemas began to appear in the comic plays of satirists such as Molière.

Enemas have also been used for other purposes. In the Middle Ages, feeble patients were given liquid nutrient enemas, to feed them if they were too weak to eat. However, it's not clear how well this would actually work. Enemas are also sometimes used to administer drugs. If a drug is likely to upset the stomach, for example, doctors may instead use an enema. In the eighteenth century, there was a brief trend for tobacco smoke enemas, which were thought to be a good way to revive victims of drowning. The practice declined after 1811, when it became clear that the tobacco smoke was toxic. Although *clyster* is now an outdated term, enemas are still quite fashionable among certain people, although today the procedure is known by the trendy-sounding name of "colonic irrigation."

what is a hemiglossectomy?

This was the name for an astonishingly brutal eighteenth and nineteenth century cure for stuttering, which involved cutting off half the tongue. Hemiglossectomies are still sometimes performed, in cases of oral cancer, where all other treatments have failed, and so part of the tongue needs to be removed. Today these operations are of course performed under general anesthetic. This wasn't the case for most of the eighteenth and nineteenth centuries.

What makes this operation so remarkable is that it seems so unnecessarily traumatic for a condition as mild as a stammer, which can be treated in many other ways. According to records, the hemiglossectomy didn't even work, and many patients bled to death as a result. In 1937, one correspondent to the *British Medical Journal* summed up the objections to the treatment:

> *There is no logical reason why it should not correct the stammering habit, but the adoption of such a course indicates a complete ignorance of speech training. The hangman can cure indigestion, but there are other and better methods!*

what causes warts? and is there a cure?

Warts are caused by the human papilloma virus, which has more than a hundred different strains. Strains 1, 2, and 3 cause the most common warts, while there are a number of other strains that cause genital warts. Warts are mildly contagious, and can be passed from person to person, as well as via indirect contact through towels, shoes, and wet floors. Warts are often found on the hands and fingers, but they can grow all over the body, including on the eyelids, rectum, or even the inside of the mouth.

Warts will sometimes heal themselves without being treated, but it's generally sensible to get them seen to. There are a number of treatments, the most effective of which is to apply salicylic acid. Warts can also be frozen with liquid nitrogen, cauterized, or removed with a laser. However, none of these treatments is thought to work more than about 75 percent of the time, and warts will sometimes grow back after being removed.

Warts have given rise to a great number of folk cures and theories over the years, no doubt because they seem to appear and then disappear for no obvious reason. The length of time a wart lasts is also very unpredictable compared with many illnesses, which is likely to have encouraged people to believe that their cure has worked if it happens to roughly coincide with the wart disappearing. By contrast, we all know that colds tend to last around seven to ten days, so when we do recover from a cold after a week or so, we are probably less likely to attribute the healing to having touched a worm or buried a pebble.

Warts are of course traditionally associated with toads, because toads have a warty appearance (although the bumps on a toad's back are not actually warts). One common theory was that you caught warts by touching a toad. To cure the wart, you would wear a live toad in a bag around your neck, until the poor creature died. Alternatively, you might rub a frog or snail on the wart, and then impale it on a twig (the frog or snail, that is). For those looking for a bloodier brand of animal cruelty, you could chop off the head of an eel, drip the blood onto the wart, and then bury the head.

There were also some rather less gruesome folk remedies out there. One theory held that a wart should be rubbed with a piece of elder wood, which was then thrown away. If there was no elder wood to hand, a bean pod would do just as well. You could also sell, or give away, your wart. To pass it on to the dead, you would have to throw a stone after a funeral procession (a risky choice, surely), or alternatively rub the wart while a funeral procession passed, or collect some mud from the mourners' shoes and apply it to the wart. Warts could also be sold, for a nominal price, or taken to a crossroads and symbolically buried. Alternatively, a sufferer could try making a face in the mirror at midnight, or blowing on the wart nine times during a full moon. In short, pretty much any old nonsense would do, which does at least make warts an instructive example of how people will come to believe in almost any remedy, as long as it is occasionally seen to roughly coincide with the odd person getting better.

why has arsenic been so widely used in medicine?

Arsenic is one of the world's most notorious and widely used poisons. Nero secured control of the Roman Empire by poisoning his stepbrother Britannicus with arsenic. In fifteenth-century Italy, the Borgias used arsenic to murder their political opponents. In seventeenth-century France, its usage was so common that it became known as *poudre de succession*, which means "inheritance powder."

Arsenic is an ideal poison because it is odorless, tasteless, and easily assimilated into food and drink. Until the nineteenth century, there was no test for its presence in the body, and its symptoms resemble those of cholera, which meant deaths would be unlikely to raise suspicion. Suspected victims of arsenic poisoning include Napoleon Bonaparte, King George III of England, and Francesco I de' Medici, Grand Duke of Tuscany.

The effects of arsenic poisoning are varied and unpleasant. One clue is that victims of arsenic poisoning will often have a garlicky smell to their breath or sweat. Mild symptoms may include a skin rash, hair loss, and white lines on the fingernails. More severe cases will suffer bloody vomiting, abdominal pain, severe diarrhea, and dehydration. One form of arsenic poisoning will even turn the patient's urine black.

For centuries, however, arsenic has also been used as a medicine. In traditional Chinese medicine, it is known as *pi shuang*, and is still used to treat cancer. Arsenic was a key ingredient in numerous patent medicines, including Fowler's Solution, a very popular tonic which was prescribed from the late eighteenth century until as recently as the late 1950s, as a cure for malaria and syphilis. Donovan's Solution, containing arsenic, was used to treat rheumatism, arthritis, and diabetes. Victorian women also used it as a cosmetic on their arms and face, to improve their complexion.

Since arsenic was well known to be poisonous, why was it used so often as a medicine? How could anyone think that arsenic would be good for them? First, one reason may be that mild arsenic poisoning breaks down the blood vessels in the face, thus appearing to give the patient a healthy, reinvigorated glow. Second, the idea of a poison also being a medicine is not as daft as it might initially sound. We naturally tend to think of poison and medicine as being two distinct, opposing concepts, but in fact, they are often one and the same. Many medicines work by attacking or poisoning harmful agents in the body, such as parasites and bacteria. In many cases, these medicines are also somewhat toxic to us too, but generally the benefits outweigh the risks.

With this in mind, it is not entirely surprising that arsenic has been used in many medicines that have actually proven to be effective, some of which are even still used today. After its discovery in 1909, salvarsan became the most widely prescribed drug in the world, as at the time it was the most effective treatment for syphilis (although it was supplanted in the 1940s by penicillin). Arsenic is still used today in chemotherapy, to treat certain forms of leukemia. It may also turn out to be effective in treating autoimmune diseases.

chapter five

the good doctor

My doctor gave me two weeks to live.
I hope they're in August.

RITA RUDNER

how did medical student vesalius get his big break?

In 1536, the twenty-two-year-old Andreas Vesalius was a lowly medical student at the University of Louvain in his native Belgium. He had just returned from studying at the University of Paris, and he found the standard of teaching at Louvain pitifully inadequate. On one occasion, seeing his lecturer's dismal attempt at dissecting an animal, Vesalius strode to the front of the class and showed him how it ought to be done.

What may have been the defining event of Vesalius's life took place soon after. Outside the city walls, he found the body of an executed criminal swinging from a gallows. The skeleton was intact, and held together by ligaments. Excitedly, he tore off the arms and legs and rushed home with them. He returned later that night, breaking the city curfew, to climb the gibbet, smash the chain, and remove the rest of the body. He then boiled up the bones, and used them to create his first skeleton. When anyone asked, he lied and said that he had brought the bones with him from Paris.

At this time in Northern Europe, the idea of dissecting a human body was practically unthinkable, which meant the study of anatomy could make only limited progress, as much of the received wisdom could never be tested and questioned. An anatomy lecture at the time consisted of studying the ancient texts of Aristotle and Galen, perhaps illustrated by a dissection of an animal such as a dog or pig, on the assumption that all animals' bodies and organs were basically the same. The focus of the lecture was the book, which was the true source of knowledge; the dissection was merely a sideshow.

Vesalius's revolutionary approach was to question the hallowed teachings of Galen, and conduct his own investigations. In doing so, he found that the great Galen had actually made numerous errors. Galen believed the human jaw consisted of two bones, but Vesalius found it was just one. Galen believed the breastbone contained seven bones, but Vesalius found just three. Galen contended that women had one more rib than men, but Vesalius couldn't find it. In total, Vesalius detailed more than three hundred such errors, and came to the conclusion that Galen had never actually dissected a human body; instead Galen must have simply dissected other animals, such as dogs, pigs, and monkeys, and assumed that human bodies were the same (it seems likely that this is indeed what Galen had done).

Vesalius's anatomy lectures became a sensation, drawing large crowds of spectators. As interest in his work grew, he was granted the corpses of executed criminals to dissect. Often, he would dissect a human corpse side by side with a live animal, which he would vivisect simultaneously. The idea was to demonstrate the action and function of each part of the body as far as possible in the live animal, while showing the equivalent part of the human corpse, and comparing how they differed. Vesalius seems to have taken a rather gruesome pleasure from vivisection. In one example, he describes his satisfaction at the vivisection of a pregnant bitch, as he could clearly see the unborn puppies struggling to breathe when the placental blood supply was cut off.

Today, Vesalius is regarded as the father of anatomy, and one of the most important figures in the history of medicine. His illustrated guide to anatomy, *De Humani Corporis Fabrica*, was enormously influential, and he made a vast number of important discoveries, particularly concerning the structure and function of the heart and circulatory system. But just as important as these insights was Vesalius's scientific method, in which experimental evidence was given more weight than the authority of the great figures of the past.

Vesalius's practices generated considerable controversy throughout his career. At the age of thirty, he was forced to give up anatomy, and became the court physician in Madrid. According to some reports, Vesalius was then banished by the Spanish Inquisition after he was found dissecting the body of a Spanish nobleman, who then woke up during the procedure. As a punishment for this outrage, he was sent on a pilgrimage to Jerusalem, but his ship wrecked, and he died of hunger on the Greek island of Zante.

In truth, Vesalius does not seem to have always been particularly rigorous in making sure his subjects were actually dead. On a separate occasion, he described removing the still-beating heart of someone who had died in an accident. We might wonder quite what Vesalius's definition of "dead" was, if a beating heart didn't constitute life!

why was edward jenner rejected by the royal college of physicians?

Smallpox has been one of the biggest killers throughout human history. No cure has ever been found, and the disease has consistently killed around 30 to 50 percent of all those who became infected. Smallpox is highly contagious, and can be easily transmitted by face-to-face contact. During the eighteenth century, it was responsible for one in every ten deaths in Europe, killing around 60 million people.

Smallpox is a deeply unpleasant disease. Within days of becoming infected, victims develop hundreds of disfiguring blisters, particularly on their face, neck, and hands. Fever and blindness often follow, with death usually occurring within ten to sixteen days.

For centuries, people had observed that anyone who survived a bout of smallpox would then
become immune. During the early eighteenth century, inoculation was introduced to Europe from
Turkey, where it had been widely practiced for more
than a century. Inoculation involved taking pus from
a person infected with smallpox, and scratching it
into the skin of the person being inoculated. Generally, the recipient contracted a much-weakened form
of the disease, with no facial scarring, and hence-
forth became immune. However, around 2 to 3 percent of those inoculated developed the full form of
the disease and died, with many also becoming the
source of fresh epidemics. As a result, there was considerable hostility to inoculation.

Some years later, an English country doctor
named Edward Jenner became aware of the local
folk wisdom which held that milkmaids in his
native Gloucestershire, who were renowned for
their beauty and clear complexions, could not catch
smallpox. The reason, they explained, was that most
of them would first catch cowpox from the cows
they worked with every day, and they were therefore
somehow protected from smallpox. Jenner decided
to conduct an experiment. He took an eight-year-old
boy named James Phipps, and infected him with
the pus from a cowpox pustule. The boy fell mildly
ill, as expected. After the boy recovered, Jenner
tried to inoculate him with smallpox, but the infection wouldn't take. Phipps was, just as Jenner had

hoped, immune to smallpox. Jenner named this procedure "vaccination," after the Latin word *vacca*, meaning "cow."

There was a great deal of opposition to vaccination, as it was seen by many as a disgusting and unnatural practice to infect people with a disease from a cow. Jenner's first paper to the Royal Society was rejected, and he had to perform a great number of further experiments before his argument was taken seriously. But the clear evidence for the effectiveness of vaccination won out, and it soon became widely used. In 1840, the British government banned smallpox inoculation, clearly demonstrating that vaccination had won the day.

In fact, Jenner did not actually invent vaccination, as it had been performed before by other country doctors (although it's not clear whether or not he was aware of this), but his success in popularizing it and proving the case mean that Jenner is widely held to have saved the lives of more people than any other man in history. In recognition of his achievement, he was appointed Physician Extraordinary to King George IV, a considerable honor, and feted and celebrated all around the world. In the 1680s, somewhere between 7 and 14 percent of deaths in London had been attributable to smallpox; by 1850, this had fallen to just 1 percent.

However, the world of medicine was harder to impress. When Jenner applied to join the Royal college of Physicians in London, he was denied entry, and told that he would first have to pass a test on the theories of Hippocrates and Galen. Jenner refused to take the test, as he felt that his achievement in defeating smallpox was sufficient to make him worthy of election. This argument was clearly not acceptable to the hidebound dinosaurs of the College, and Jenner was thus never accepted into it.

why was hungarian doctor ignaz semmelweis hounded out of medicine?

Ignaz Semmelweis worked in the maternity wards of the Vienna General Hospital in the 1840s, and was very troubled by the wards' terrible fatality rates from puerperal fever, also known as "childbed fever." The hospital had two obstetrical clinics: one was run by male medical students, while the other employed midwives. It was well known that the death rate in the students' clinic was much higher than that in the midwives' clinic, but no one knew why. Pregnant women would literally beg to be admitted to the midwives' clinic, and many simply chose to give birth in the street, rather than take the risk of being treated in the students' clinic, where mortality rates were as high as 15 percent.

Semmelweis was determined to work out why the mortality rates in the two clinics were so divergent, but he could find no significant difference between the two clinics, apart from the staff they employed. In 1847, Semmelweis's friend Jakob Kolletschka died from a condition similar to puerperal fever after being accidentally cut by a scalpel while performing an autopsy. This was the breakthrough. Semmelweis realized that the medical students were somehow becoming contaminated by the corpses upon which they were performing autopsies, and were then transferring this contamination to the expectant mothers in the obstetrical clinic when they went straight to the delivery rooms. Midwives did not perform autopsies, which is why the midwives' clinic was not contaminated in this way.

Semmelweis instituted a radical new rule: All medical staff were now to wash their hands and instruments in chlorinated lime solution in between performing autopsies and dealing with patients. The results were immediate, as the mortality rate in the doctors' clinic quickly dropped to the level of that in the midwives' clinic. It was a remarkable success.

But Semmelweis's discovery proved extremely controversial, for a number of reasons. First, the mainstream medical view was that illnesses were caused by an imbalance in the four humors, which was unique to the patient. This thinking argued that all conditions were individual and specific to the patient, whereas Semmelweis was arguing that one single, invisible cause was responsible for all the deaths. Second, diseases were believed to be spread by "bad air" or miasmas; there was no known mechanism for them being carried in tiny amounts on a person's hands. Third, Semmelweis's theory carried the implicit suggestion that doctors were responsible for spreading disease, rather than curing it, which was a heretical notion. As a result, Semmelweis was derided and criticized by the medical establishment, and hounded out of Vienna.

He moved to Pest (now Budapest), and in 1851 took an unpaid role at the obstetric ward at the small St. Rochus hospital. When he arrived, childbed fever was out of control at the hospital, but Semmelweis quickly wiped it out—between 1851 and 1855, less than 1 percent of the hospital's patients died of the disease. Nonetheless, Semmelweis's success was still overlooked by the medical establishment, and his methods were ignored.

He became angry and bitter, writing frequent polemics condemning his critics. As his health deteriorated, and his behavior became more erratic, he was committed to a lunatic asylum. He died two weeks later, after being severely beaten by the guards and kept in a straitjacket in a darkened cell. He died of blood poisoning, a condition that in a maternity ward would be called childbed fever. The rules of the Hungarian Association of Physicians required that each member who died should be honored with a commemorative address, but

Semmelweis's death was ignored, and no colleagues attended his funeral.

Since his death, however, Semmelweis's contribution to medicine has become recognized, and today he is regarded as a pioneer of antiseptic medicine, and many medical institutions in Austria and Hungary now bear his name.

did an african-american pioneer of blood transfusions die because he was refused a blood transfusion, on account of the color of his skin?

Charles Drew was an American surgeon who was born in Washington, D.C., in 1904. He was a leading researcher in the field of blood transfusions, and in the course of his career he developed new techniques for the storage of blood, which were key to the development of large-scale blood banks that saved many thousands of lives during World War II. Drew also took an active role in this project; as the director of the "Blood for Britain" program, he supervised the collection of 14,500 pints (6,861 liters) of blood plasma for the British. Drew also campaigned against the contemporary practice of keeping the blood of African-Americans separate from that of white donors, a practice which was based on prejudice, and had no scientific merit.

In 1943, Drew became the first African-American elected to serve as an examiner on the American Board of Surgery.

At around eight a.m. on April 1, 1950, Drew was driving back from Tuskegee, Alabama, having spent the night working at the free clinic. No doubt exhausted after his night's work, Drew crashed his car, injuring himself along with the three colleagues who were traveling with him. The three other doctors suffered minor injuries, while Drew was trapped in the wreckage, with his foot wedged under the brake pedal. When the emergency services arrived, Drew was in shock, with severe leg injuries. He was taken to Alamance General Hospital in Burlington, North Carolina, where he was pronounced dead.

There is a recurring version of this story, which states that Drew was denied care because of the color of his skin. Specifically, this version of the story holds that this pioneer of blood transfusions was, in a bitterly ironic twist, himself denied a blood transfusion, and consequently bled to death. It is a powerful story that highlights the shameful practice of segregating hospitals that did go on in the South at that time.

Despite its dramatic power, the story appears to be untrue. According to one of Drew's passengers that night, Dr. John Ford, "We all received the very best of care that night. The doctors started treating us immediately." In fact, according to Ford, Drew's injuries were such that a blood transfusion would have been actively harmful. "He had superior vena cava syndrome: blood was blocked getting back to his heart from his brain and upper extremities," Ford said. "To give him a transfusion would have killed him sooner. Even the most heroic efforts couldn't have saved him. I can truthfully say that no efforts were spared in the treatment of Dr. Drew, and, contrary to popular myth, the fact that he was a Negro did not in any way limit the care that was given to him."

This story is similar to another myth, concerning the legendary blues singer Bessie Smith. The story goes that Smith died after a car accident, when a "whites only" hospital refused to treat her. In fact, it seems that the scene of Bessie Smith's accident was served by two ambulances, one from the black hospital for Smith, and the other from a white hospital. Smith was taken to Clarksdale's Afro-American Hospital, where she died. Although it seems entirely plausible that she might have been refused treatment if she had been taken to the white hospital, this never happened. Bessie Smith did receive treatment at the Clarksdale Hospital, including having her right arm amputated, but her injuries were severe, and as a result she died in the hospital.

who was the real sherlock holmes?

Before becoming a celebrated writer, Arthur Conan Doyle studied medicine at Edinburgh University. One of his lecturers there was the brilliant Dr. Joseph Bell, who would prove to be a pivotal influence on the young Conan Doyle. In his second year, Conan Doyle was chosen to serve as Bell's outpatient clerk at the Edinburgh Royal Infirmary, and so got to observe his methods at close hand. Bell was tall and lean, with piercing green eyes and a large, hooked nose.

Bell was a brilliant diagnostician who would make amazing deductions about people just by observing their accent, clothing, gait, complexion, and other small details. In this way, he impressed upon his students the importance of close observation in making a diagnosis. On one occasion, a patient entered the room, and Bell immediately observed that he was a recently discharged non-commissioned officer who had served in a Highland regiment stationed in Barbados. He explained his deductions: "You see, gentlemen, the man was a respectful man but did not remove his hat. They do not in the army, but he would have learned civilian ways had he been long discharged. He has an air of authority and is obviously Scottish. As to Barbados, his complaint is elephantiasis, which is West Indian and not British."

Similar moments of deductive brilliance are of course a recurring feature of Conan Doyle's eternally popular Sherlock Holmes stories. In "A Study in Scarlet," we see the first meeting between Holmes and his assistant Watson, which has obvious echoes of the scene just described. Upon meeting Watson, Holmes observes, "You have been in Afghanistan, I perceive." Watson assumes someone has told Holmes about him, but the detective explains:

> *"Nothing of the sort. I knew you came from Afghanistan. From long habit the train of thoughts ran so swiftly through my mind that I arrived at the conclusion without being conscious of intermediate steps. There were such steps, however. The train of reasoning ran, 'Here is a gentleman of a medical type, but with the air of a military man. Clearly an army doctor, then. He has just come from the tropics, for his face is dark, and that is not the natural tint of his skin, for his wrists are fair. He has undergone hardship and sickness, as his haggard face says clearly. His left arm has been injured. He holds it in a stiff and unnatural manner. Where in the tropics could an English army doctor have seen much hardship and*

got his arm wounded? Clearly in Afghan-
istan.' The whole train of thought did not
occupy a second. I then remarked that you
came from Afghanistan, and you were as-
tonished."

The dull-witted Watson replies, "It is simple enough as you explain it."

Correspondence from Conan Doyle confirms that Bell was the inspiration for Sherlock Holmes. After his creation had brought him fame and success, Conan Doyle wrote to his former teacher, "It is most certainly to you that I owe Sherlock Holmes, and though in the stories I have the advantage of being able to place him in all sorts of dramatic positions, I do not think that his analytical work is in the least an exaggeration of some effects which I have seen you produce in the out-patient ward." Bell generously replied, "You are yourself Sherlock Holmes and well you know it."

who discovered antisepsis?

Joseph Lister (1827–1912) was an English surgeon who pioneered the concept of sterile surgery. While working at the Glasgow Royal Infirmary, Lister instigated a policy of antiseptic surgery that involved spraying the operating theater with carbolic acid, and cleaning the instruments, dressings, and even the wounds themselves with the same solution. Thanks to Lister's sanitary innovations, the rates of postoperative infection and mortality dropped markedly, and his ideas were widely adopted.

Various forms of antisepsis had been used since ancient times, before germ theory was ever discovered. Ancient man would hang meat to dry in the sunlight, or over a slow smoky fire, as this meant the meat would resist decay much longer. The ancient Egyptians preserved their mummies using smoke, dry tombs, and embalming chemicals. Early medicine in a number of cultures included the application of naturally occurring antiseptic tars and petroleums onto wounds and sores. Around 800 B.C., the Hindu surgeon Susruta recommended fumigating operating theaters before and after surgery. He also recommended that water could be kept clean by being stored in copper vessels, filtered through charcoal, and exposed to heat. Nonetheless, although certain antiseptic techniques may have been used before Lister, his reputation as the father of antisepsis is well deserved, as it was he who first applied Pasteur's discoveries of germ theory to the practical improvement of surgery.

While Lister's approach was revolutionary at the time, if we saw his operating theater today it would not strike us as being particularly clean at all. Although he endeavored to keep the surgical area itself clean, Lister's operating theaters were not much cleaner than the rest of the hospital. Lister himself performed surgery using the same apron every time, which was so caked in blood that it shone. Today, surgery follows the principle of asepsis rather than antisepsis, meaning that rather than simply killing germs during the surgical procedure (antisepsis), the aim is to make the operating theater and instruments completely free of germs in the first place (asepsis).

what was the guinea pig club?

The Guinea Pig Club was the name given to a group of burn victims who underwent treatment together at the Queen Victoria Hospital in East Grinstead, England, during World War II. Most of the patients were injured fighter pilots and bombers who had all suffered severe burns, often to the face and hands. The program was run by Archibald McIndoe, who was later knighted for his efforts. They were known as the Guinea Pig Club because many of the reconstructive plastic surgery techniques that were used were new and experimental. Before the war, patients had usually died from severe burns, and there were no precedents for much of the work that McIndoe carried out.

The group was formed in 1941, and consisted of thirty-nine patients. To join the club, you had to have gone through at least ten surgical procedures. Many of the victims were badly disfigured.

Air gunner Les Wilkins had lost his face and hands, and McIndoe had to make incisions between his knuckles to re-create fingers.

McIndoe is regarded as a pioneer in the field of plastic surgery, as his innovations and inventions marked an important advance in the subject. The Guinea Pig Club was also notable for McIndoe's unusual approach to patient care and rehabilitation. Knowing that many of the patients would be at the hospital for years, he tried to make the conditions as relaxed and communal as possible. Patients were not required to wear the standard "convalescent blues," and they could leave the hospital at any time. Barrels of beer were kept in the wards, and McIndoe encouraged local families to take in the patients as guests. Burn victims are often badly disfigured, and dealing with the reactions of strangers is a major challenge, but East Grinstead became known as "the town that didn't stare."

The sense of humor and solidarity that was fostered is evident in many details. When appointing board members, the club chose a secretary who had suffered badly burned fingers, so that the minutes would not be too verbose. The club's first treasurer was an airman who had severe burns to his legs, which meant he would not be able to run away with the club's funds. There is even a lighthearted club song, whose first verse goes like this:

We are McIndoe's army,
We are his Guinea Pigs.
With dermatomes and pedicles,
Glass eyes, false teeth and wigs.
And when we get our discharge
We'll shout with all our might:
"Per ardua ad astra"
We'd rather drink than fight!

After the war ended, the group continued to meet socially, and by this point it counted more than six hundred members. The last of their annual get-togethers was in 2007, but because of the age and frailty of the surviving members, there appear to be no plans for any further meetings.

how did a single london water pump change the course of modern medicine?

Cholera was the scourge of the nineteenth century. It was initially endemic to India, where it was believed to be spread by the contaminated Ganges River, but from 1816 it began to spread along trade routes across Asia. This first cholera pandemic lasted a decade, until 1826, and almost reached Europe before receding. The second pandemic started in 1829, and spread across Asia before reaching Egypt, North Africa, Russia, and then finally Western Europe. Further pandemics continued throughout the nineteenth century and beyond, causing an estimated 38 million deaths in India alone.

Cholera was a horrific way to die. Sufferers would develop acute nausea, which would develop into violent vomiting and diarrhea. Their stools would become a gray liquid, described as "rice water," which soon consisted of nothing more than liquid and fragments of gut. This would lead to extreme, painful cramps, and an insatiable thirst. As the patient neared death, he would develop the classic mark of cholera: pinched blue lips in a shriveled, sunken face.

No one knew what caused the disease, or how it spread. In London, there had been a number of epidemics since 1832, resulting in tens of thousands of deaths. In 1854, there was a severe outbreak of cholera in a small area of Soho, around Broad Street (which is now renamed Broadwick Street). John Snow was a physician based in London who was skeptical of the prevailing "miasma" theory, which held that diseases were spread by "bad air." Snow studied the Soho outbreak, and observed that the vast majority of cases had occurred in homes near just one water pump, on Broad Street. Furthermore, there were only ten deaths in homes which were nearer to a different water pump, but Snow learned that most of the victims in these homes actually took their water from the Broad Street pump, either because they preferred it, or because they were children whose school was near the pump.

Snow examined the pump and its water closely, but although he couldn't physically identify the specific cause of the outbreak, his statistical and geographical analysis was enough to convince the local council to disable the pump by removing the handle, and this measure alone succeeded in ending the outbreak. It was later discovered that the well supplying the Broad Street pump had been dug just three feet from a leaking cesspit that had become infected with cholera.

Snow's intervention had powerfully demonstrated that diseases could be transmitted through fecally contaminated water, rather than through miasmas. Although his findings were not immediately accepted by the scientific community, Snow's analysis of the Broad Street outbreak is now regarded as the defining event in the science of epidemiology. Today, there is still a replica water pump on Broad Street to mark Snow's achievement, just yards from the pub named in his honor.

what was unusual about dr. james barry?

James Barry was a leading surgeon in the British army in the early to mid-nineteenth century. He graduated from Edinburgh University, and served in Canada, India, and South Africa, reaching the rank of inspector general. He was known as a passionate reformer who helped to improve the lives of both soldiers and civilians. He campaigned for better food and care for lepers and prisoners, and in 1826 became the first British surgeon to perform a successful cesarean section in which both mother and child survived. Barry was a fiery character who fought a number of duels, and was arrested and disciplined on a number of occasions. Florence Nightingale described Barry as "the most hardened creature I ever met throughout the army," after they clashed in Corfu during the Crimean War.

In 1865, Barry fell victim to an epidemic of dysentery, and died in London. While preparing the body for burial, housemaid Sophia Bishop made an astonishing discovery. According to Bishop, Barry was actually "a perfect woman!" Not only that, but a woman who had given birth, to judge from the stretch marks Bishop observed on the body. This meant that Barry had been the first woman to qualify as a doctor in Britain, and that she had successfully hidden her gender from her staff and colleagues for almost fifty years.

However, it seems that some of them may have had their suspicions. When posted to South Africa in 1824, Barry was seen to have a close relationship with the colony's governor, Lord Somerset, which even led to the accusation of a homosexual relationship. Barry's colleagues had also noticed her smooth chin, small frame, and high-pitched voice. She wore padded shoulders, and shoes with lifts. Lord Albemarle remarked that Barry "showed a certain effeminacy in his manner." After Barry's death, when Bishop's claim became public, a number of Barry's colleagues claimed to have known all along.

A recent discovery of family correspondence confirms that Barry's family helped her to pull off an elaborate conspiracy, which began before her acceptance to Edinburgh University. Barry was born in Cork some time between 1792 and 1795 as Margaret Bulkley, and the death of her father left the family destitute. Having moved to London, Margaret and her mother came up with a plan that would allow her to fulfill her considerable potential while financially supporting the pair of them. This was of course a time when only men were allowed to study medicine and become doctors. Using their influential friends, Margaret began privately studying medicine. In 1809, the pair set off on a voyage for Scotland. Before boarding the boat, Margaret changed her identity, presenting herself as James Barry throughout the trip, and thus avoiding any risk of discovery on their arrival. From that point on, she seems to have lived her life as a man, and escaped discovery until her death.

was a derided female scientist the true discoverer of the structure of dna?

The woman in question was Rosalind Franklin, and although there is still considerable debate about exactly how much of the credit she deserved for this momentous breakthrough, few would dispute that she played a crucial and undervalued role. Franklin was a brilliant biologist and chemist who had learned X-ray diffraction techniques at the Laboratoire Central des Services Chimiques de l'État in Paris.

In 1951, she joined the department at King's College, London, which at that time had a patronizing and sexist culture toward women. Only men were allowed in the university dining rooms, and after hours Franklin's colleagues went to men-only pubs. Franklin was assigned to lead a research project into the structure of DNA. Her colleague Maurice Wilkins was working on a similar, separate project. They were peers, but Wilkins automatically assumed that Franklin was his technical assistant, which made their relationship frosty from the start.

By this time, much was already known about DNA and its role as a store of genetic material. What was not known was what the molecules looked like, or how they functioned. Franklin used pioneering crystallographic X-ray photography to take incredible pictures of the intricate double helix of DNA, which have been described as "the most beautiful X-ray photographs of any substance ever taken."

Without Franklin's knowledge, Wilkins showed her work to James Watson, who was pursuing similar research into DNA at the University of Cambridge. Franklin's research, and in particular her photographs, provided Watson with the crucial data he needed to complete his own model of DNA. Watson quickly published his discovery in the scientific journal *Nature*, naming only himself and his colleague Francis Crick as authors, with only a vague footnote mentioning Franklin and Wilkins's work. The research carried out by Wilkins and Franklin was also included separately, with Wilkins's paper coming second, and Franklin's third.

Franklin left Kings, and continued to carry out groundbreaking research, including work on the tobacco mosaic virus and then polio. In 1956, she became ill with cancer, and died. It is possible that her cancer may have been caused by exposure to radiation while working on the X-ray techniques that revealed the structure of DNA. After her death, James Watson released a book called *The Double Helix*, in which Franklin was mentioned only fleetingly, and in disparaging terms. In the book, Watson describes Franklin as being Wilkins's assistant, and suggests that she was incapable of interpreting her own data. In 1962, when Watson, Crick, and

Wilkins were awarded the Nobel Prize, Franklin wasn't even mentioned.

Today, her contribution is increasingly recognized. Her likeness at the National Portrait Gallery in London hangs alongside those of Watson, Crick, and Wilkins, and in 1992 English Heritage finally placed a blue plaque on her childhood home. In 2003, the Royal Society established the Rosalind Franklin Award, to reward an outstanding contribution to any area of natural science, engineering, or technology. In 2009, the readers of *New Scientist* magazine voted Franklin second only to Marie Curie in a poll to find the most inspirational female scientist.

why did baron dupuytren take fat from corpses?

Baron Guillaume Dupuytren (1777–1835) was a leading French surgeon who was celebrated for his skill and dexterity. He treated Napoleon Bonaparte's hemorrhoids, and he was one of the first surgeons to successfully drain a brain abscess using trepanation, in which a hole is made in the skull. He was a prolific writer, and over the course of his life he amassed a great fortune.

He is perhaps best remembered today for Dupuytren's contracture, a condition in which the fingers bend inward toward the palm and cannot be straightened. Dupuytren was the first person to devise and perform an operation to correct this condition, which is ironic given that he himself was so tight-fisted. According to legend, when he was a medical student, Dupuytren would take fat from corpses in the dissection room to burn in his reading lamp at home. Dupuytren was widely admired, but equally disliked, as he was seen to be haughty, unfriendly, and obsessive.

On one occasion, he had saved the life of a duchess. The grateful lady presented him with a beautiful, hand-embroidered purse as a token of her gratitude, but Dupuytren snapped back that his fee was 5,000 francs. The duchess took back the purse, opened it, removed five 1,000 franc notes, and then handed it back to him, telling him that it now contained exactly the amount he'd asked. Smiling, she complimented him on his modest fees.

chapter six

bad medicine

Never go to a doctor whose office plants have died.

ERMA BOMBECK

was jack the ripper a surgeon?

Jack the Ripper was one of the world's most infamous serial killers, whose unsolved series of murders make up one of the most mysterious and fascinating cases in the annals of crime. He (or, according to some, she) is generally believed to have committed somewhere between five and eleven murders between 1888 and 1891, in the impoverished slums of Whitechapel, in the East End of London. But there is considerable doubt and debate about almost every detail of these crimes.

If there can be said to be a consensus, it is that Jack the Ripper committed five murders, known as the "canonical five," between August and November 1888. The victims were Mary Ann Nichols (killed August 31), Annie Chapman (September 8), Elizabeth Stride and Catherine Eddowes (both September 30), and Mary Jane Kelly (November 9). All five women worked as prostitutes, and they were all killed by having their throat cut, before being horribly mutilated (with one exception—Elizabeth Stride—it is believed that Jack was interrupted and fled).

The murders remain unsolved, despite one of the largest police investigations in history, as well as the ongoing efforts of an army of journalists, academics, and amateur investigators, who have turned "Ripperology" into a field of study all of its own. So far, more than a hundred people have been identified by Ripperologists as potential suspects, even though most of them were never considered credible by the police at the time, and there are a whole range of fascinating conspiracy theories.

However, for the purposes of this question, the interesting detail concerns the Ripper's medical knowledge. At the time, many experts, including the police, were convinced that the Ripper must have been a butcher, an abattoir worker, or a doctor of some kind, because of the speed and skill with which he attacked and dismembered his victims. Four of the five victims had their abdomen cut open. The killer removed the uterus of Annie Chapman, the heart of Mary Jane Kelly, and the uterus and left kidney of Catherine Eddowes—half of the kidney was later posted to the police, inside one of a series of taunting letters, with the sender claiming to have eaten the other half. Witnesses claimed to have seen a "shabby, genteel man" carrying a shiny black bag at the scene of one of the murders, which added to the suspicion that the killer was a medical man. The police interviewed all the local butchers and abattoir workers, but found they all had alibis, eliminating them from the investigation. As a result, a number of doctors have long been considered key suspects. They include the following:

Dr. Francis J. Tumblety was a fifty-six-year-old American quack who had toured the United States and Canada selling "Indian" herbs, and was arrested for his involvement in the assassination of Abraham Lincoln, but never charged. Tumblety was in England in 1888, and he became one of the Metropolitan Police's four main suspects for the Ripper murders. He was arrested in November 1888 for "offences of gross indecency," which in this case seems to have meant homosexuality. He fled the country, and his flight coincided with the end of the killings, which adds to the case against him.

Michael Ostrog was another of the police's four central suspects. Ostrog was a Russian-born con artist. He had a record of fraud and petty theft, but nothing more serious or violent. He did claim to have been a surgeon in the Russian army.

George Chapman was born in Poland as Seweryn Kłosowski; he changed his name when he moved to England, around 1887–1888. Chapman did murder three of his wives, by poisoning; crimes for which he was hanged in 1903. He was known to be a violent misogynist, and he was also a trained surgeon. Still, it's very unusual for a serial killer to use two such differing methods, and there's little meaningful evidence to link Chapman to the specific Ripper murders.

Thomas Neill Cream was a Scottish doctor who specialized in abortions. Cream is known to have committed at least five murders, and most of his victims were prostitutes, whom he murdered by poisoning. He was thought to have been in prison at the time of the Ripper murders, but some believe he may have bribed his way out, or even arranged for a look-alike to take his place. When he was hanged in 1892, his last words on the gallows are said to have been, "I am Jack the . . ."

As the years have passed by, and the facts have become ever more distant, increasingly outrageous suspects have been suggested. They include Lewis Carroll, the writer of *Alice's Adventures in Wonderland*; Sir William Gull, physician to Queen Victoria; Prince Albert Victor, the queen's grandson; Sir John Williams, obstetrician to the queen's daughter Princess Beatrice; and the noted artist Walter Sickert. Given the amount of time that has passed, and the efforts that have already been made, this mystery will surely never be solved, but it still makes for a fascinating puzzle.

which renowned surgeon accidentally cut off his patient's testicle?

Robert Liston was a leading surgeon of the nineteenth century who was famous for the extraordinary skill and speed with which he could perform complex surgery. In an era before anesthetics, the speed of an operation was arguably the most important factor in terms of pain and the patient's chances of survival. It was said that Liston could amputate a leg and sew up the end in just 90 seconds. Liston was tall and macho; he would stride into the operating theater and instruct those in the gallery to time him with their pocket watches as he set to work. To free his hands, he would hold the bloody knife between his teeth. According to his admirers, "the gleam of his knife was followed so instantaneously by the sounds of sawing as to make the two actions appear almost simultaneous."

But this emphasis on speed could sometimes come at the expense of accuracy. On one occasion, Liston amputated a patient's leg in just two and a half minutes, and also accidentally sliced off the patient's left testicle. The audience for another amputation saw Liston cut off two of his assistant's fingers, as well as the coattails of a distinguished spectator. The spectator dropped dead on the spot from sheer terror. Later, both the patient and Liston's assistant died from gangrene. This extraordinary event has been memorably described by the great medical historian Richard Gordon as, "The only operation in history with a 300 percent mortality."

what were "ether frolics"?

The history of anesthesia is rather odd in many ways. When they were initially discovered in the eighteenth and nineteenth centuries, the first effective anesthetics were used solely for entertainment, and it barely seemed to occur to anyone to use them to alleviate the dreadful pain of surgery.

In the past, numerous methods of pain relief had been tried, with little success. In ancient Egypt, patients would be hit on the head with a mallet. Later, Europe's doctors spent centuries experimenting with ice, snow, hypnotism, suffocation, sleeping draughts, opium, wine, and cannabis. Baron Dupuytren, whom we also encountered in the previous chapter, developed the innovation of shocking his female patients with crude insults, in the hope of making them faint. But none of these solutions was satisfactory. The only things that worked were the drugs, but the problem with drugs, as Fallopius complained, was that "when soporifics are weak, they are useless, and when strong, they kill." Because of this risk, a seventeenth-century French law had banned the use of any painkillers during surgery.

Nitrous oxide, also known as laughing gas, was discovered by Joseph Priestly in 1772, but it was used primarily to create entertaining stage shows, which would involve members of the audience coming on stage, inhaling a dose from an Indian-rubber bag, and then laughing, dancing, tumbling, and generally going berserk for the amusement of the crowd. A dentist named Horace Wells did notice the drug's painkilling potential, but after one unsuccessful public demonstration he gave up and lost interest.

Ether was first synthesized in 1540, and its stupefying powers were well known. By the early nineteenth century, one of the more popular pastimes in fashionable society was to hold "ether frolics," which were parties fueled not by champagne and cocktails, but by ether. The parties were a hit, but it didn't seem to occur to anyone to use ether to relieve pain until the 1840s, when an American doctor called Crawford Long used it to painlessly remove a cyst from a friend's neck.

Even after anesthetics had been convincingly shown to work, there was still considerable hostility to their use. Ether was referred to by surgeons as the "Yankee Dodge," as doctors seemed to feel that avoiding pain was somehow cheating, or unsporting. Perhaps their professional pride was stung. In one stroke, anesthesia had made much of the surgeon's craft useless, as their speed, skill, and indifference to pain were no longer of any value. Although the doctors were unconvinced, the public reaction was more forthright. The *People's London Journal* ran the headline "Hail happy hour! We have conquered pain!" and for a brief time Anesthesia was a fashionable name for newborn girls.

what did cézanne, monet, and van gogh have in common?

Well obviously they were three of the most important painters of the Impressionist and Post-Impressionist period, but fine art falls outside the scope of this book, mainly because the extent of my knowledge on that subject can be summed up in the phrase, "I like pretty pictures." For the purposes of this book, the more relevant connection is that they may all have suffered from the same ailment.

The likely cause was a type of paint, specifically a pigment called Emerald Green, which was very popular among artists of the time. Emerald Green is made from copper acetate and arsenic trioxide, and it is the arsenic in Emerald Green that is the issue, arsenic being of course extremely toxic. Emerald Green is no longer used today, but none of the modern equivalents is quite as vivid or brilliant as Emerald Green.

Emerald Green is said to have been Paul Cézanne's favorite pigment, and his paintings are full of it. Vincent van Gogh and Claude Monet also used it heavily, and all three men suffered from symptoms which could have been caused by chronic arsenic poisoning: Cézanne developed severe diabetes, van Gogh suffered from neurological disorders, and Monet went blind. All three ailments may well have been caused by their use of Emerald Green.

On the other hand, a number of other pigments might also have been the cause. Another popular pigment was a purple called Cobalt Violet, which also contained arsenic. Other common pigments contained lead, mercury, and turpentine, which are all toxic. One way or another, it does seem likely that all three men really did suffer for their art.

was king george v killed by his doctor?

The final years of King George V's life were marred by illness, thanks in part to his heavy smoking. He suffered from emphysema, pleurisy, and bronchitis, and on January 15, 1936, he took to his bed at Sandringham House, Norfolk, complaining of a cold. His condition swiftly deteriorated, and by January 20, he was close to death. His doctor, Lord Dawson, issued a brief statement: "The King's life is moving peacefully towards its close." At 11:55 p.m. that night, the king passed on.

But George V did not die of what might be called "natural causes." The cause of death was a deliberate, fatal injection of three-quarters of a gram of morphine and one gram of cocaine, administered by Lord Dawson. The reasons for this course of action seem to have been twofold. First, the king was clearly dying, and letting him struggle through his last hours was felt to be beneath his dignity, and distressing for his family. Years later, Lord Dawson explained:

> *At about 11 o'clock it was evident that the last stage might endure for many hours, unknown to the patient but little comporting with the dignity and serenity which he so richly merited and which demanded a brief final scene. Hours of waiting just for the mechanical end when all that is really life has departed only exhausts the onlookers and keeps them so strained that they cannot avail themselves of the solace of thought, communion or prayer.*

The other, scarcely believable, reason was that the royal family were anxious that the news of the king's death should appear first in the morning edition of the *Times* newspaper, rather than the "less appropriate" evening papers, or even the upstart BBC wireless. Therefore, it was decided that the king should die in time to meet the print deadline of *The Times*. Dawson had even prewarned the editor to hold the front page.

Of course, none of this became public knowledge at the time. In fact, it only came out fifty years later, in 1986, when Lord Dawson's notes were released to the public. There was surprisingly little public outcry, no doubt largely because of the length of time that had passed. A spokesman for Buckingham Palace would say only, "It happened a long time ago, and all those concerned are now dead." Yet there were a few outraged voices. George's biographer Kenneth Rose was appalled, stating that, "In my opinion, the King was murdered by Dawson."

presumably, doctors never harmed any other kings, though, did they?

I'm afraid they did, although in most cases the damage tended to result from the doctors' well-meaning incompetence, rather than deliberate euthanasia. King Charles II of England, for example, was effectively tortured to death by his doctors, who spent the last four days of his life trying every kind of barbaric treatment without success. He was repeatedly bled, purged, clystered, cupped, blistered, and sweated by his baffled physicians. In the end he died of apoplexy—an outdated term which essentially means blood loss and loss of consciousness, the very things his doctors were causing, so it was almost certainly his treatment that killed him, rather than any underlying illness.

King George III was also poorly treated by his doctors. George suffered in the last years of his life from regular bouts of mental illness, which, it is now thought, may have been caused by a blood disease called porphyria. By 1810, George was dangerously ill, practically blind, and in considerable pain. The proper treatment at the time for someone with George's symptoms of mental illness was to force the patient into a straitjacket and so, very respectfully, this is what the king's doctors did.

In June, 323 B.C., Alexander the Great developed a fever after going on a two-day drinking bender. Within twelve days he was dead, having gradually become unable to speak. There are various theories about his death, including that he was assassinated by poisoning, but the most likely explanation is that he died of an overdose of hellebore, a herbal medicine given by his doctors. His last words were said to be, "I die by the help of too many physicians."

George Washington was also almost certainly killed by his doctors. He fell ill on Friday, December 13, 1799, after having ridden around his farm in the rain the night before, and then having eaten dinner while still in his wet clothes. On Saturday, he felt worse, so he asked his plantation manager to take some of his blood; the man took about a cup and a half (300 milliliters). Over the next few hours, two doctors arrived, and each of them bled Washington—in total, he probably lost more than half his blood in a single day. He was also purged, vomited, and blistered, all of which would have dehydrated him further. By the end of the day, instead of flowing normally, Washington's blood was thick and sticky, and oozed slowly out of his veins. He died that evening, having asked his doctors, "Pray take no more trouble about me. Let me go quietly."

President James A. Garfield was also probably killed by his doctors. He was shot by assassin Charles J. Guiteau on July 2, 1881, while waiting for a train. In those days, it was not routine for presidents to have a security detail. During an eleven-week vigil, his condition fluctuated, until his eventual death on September 19. Although Garfield was shot more than once, many believe that he would have survived if it were not for his incompetent doctors, who repeatedly probed his wounds with unsterilized fingers, trying to locate a bullet that was lodged inside him. One doctor punctured the president's liver in doing so. Blood poisoning and infection developed, and Garfield lost more than 60 pounds before eventually dying of a heart attack.

is it true that some patients wake up during surgery?

Horrifyingly, this does happen. "Anesthesia awareness" is the term, and it is currently one of the major talking points in the world of anesthesia. What generally seems to happen is that patients are given a general anesthetic to make them unconscious and insensitive to pain, as well as a muscle relaxant such as curare, which paralyzes their muscles, making them unable to move or communicate in any way.

But if the equipment delivering the anesthetic is faulty, or if the anesthetist makes a mistake, the patient may regain consciousness and feel pain, but be unable to communicate this to the medical team in any way because of the muscle relaxant. Some patients may also be more resistant to anesthetics, if they are long-term users of tobacco, alcohol, or opiate drugs such as heroin. The risk of side effects grows depending on how much anesthetic is given, so anesthetists have to tread a fairly fine line, providing just enough anesthetic to prevent consciousness, while minimizing the risk of complications that come with providing too much anesthetic.

According to estimates, anesthesia awareness occurs in around 0.1 to 0.2 percent of operations. Only around 42 percent of these will actually feel pain, but the vast majority will panic and become anxious, and many develop lasting psychological effects, including post-traumatic stress disorder.

In recent years, scientists have developed a number of techniques and technologies to try to address this important problem. These include a number of different brain wave monitors, which monitor electrical activity in the brain. In theory, the rate of electrical activity is distinctly reduced when a person is under general anesthetic, but none of these systems seems to be perfect at present, largely because some anesthetics do not reduce the electrical activity rate in this way.

Another potential solution is the "isolated forearm technique," which has been pioneered by Professor Wang and Dr. Ian Russell at Hull Royal Infirmary. The technique works as follows. A tourniquet is applied to the forearm, cutting off the blood supply, just before the muscle relaxant is administered. As a result, if at any point the patient regains consciousness, and wants to communicate this to the medical team, they will be able to move their hand.

Bizarrely, an alternative solution may simply be to deliberately keep more patients awake. In recent years, doctors have conducted a number of operations under what is known as "light anesthesia" or "blocks." This technique involves an epidural injection into the spine, which blocks pain, but allows the patient to stay awake and breathe and talk normally. Using light anesthesia, doctors have been able to carry out major operations, including heart bypass surgery. Light anesthesia not only averts the risk of anesthesia awareness (as patients can simply tell the doctor if they find they are in pain), but also avoids some of the risks and side effects associated with general anesthesia. There are other benefits too. Light anesthesia generally allows patients to go home earlier, increasing patient satisfaction, and cutting costs.

Finally, did you know that people with red hair need more anesthesia than blondes and brunettes? Amazingly, there seems to be a link between one hormone that affects skin pigmentation, and another that relieves pain. As a result, natural redheads are more sensitive to certain types of pain, and consequently require around 20 percent more anesthetic to reach an equivalent level of insensitivity!

was abraham lincoln poisoned?

During the Civil War, Abraham Lincoln was noted for his calmness and composure under pressure. However, only a few years earlier, he had been regarded as a violent, volatile man, prone to flying into verbal and physical rages. During an 1858 debate, he grabbed a former aide and shook him "until his teeth chattered." According to his partner at a law firm, Lincoln could get "so angry that he looked like Lucifer in an uncontrollable rage." So what changed?

The answer may be that Lincoln was a long-term victim of self-inflicted mercury poisoning. During the 1850s, Lincoln suffered from what he described as "melancholy," which sounds like it might have been what we would today call clinical depression; one contemporary described him as living in a "cave of gloom." To remedy this, he took a drug called blue mass, which was widely used in the nineteenth century for a whole range of conditions, including depression, toothache, constipation, and the pain of childbirth.

Scientists at the Royal Society of Chemistry have recently tested genuine samples of blue mass pills, and found that their main effective ingredient was mercury, which is of course highly toxic. The pills contained more than 120 times what is now considered to be a safe daily dose of mercury. The symptoms of mercury poisoning include nausea, diarrhea, vomiting, and dehydration. It can also cause insomnia, tremors, and outbursts of rage, all of which we know Lincoln suffered from.

Soon after his inauguration in 1861, Lincoln stopped taking the pills. In a letter to a friend, he explained that they "made him cross." The symptoms of mercury poisoning are reversible, which may well explain how the violent, temperamental Lincoln became the calm, unruffled statesman of the Civil War.

which popular drug may save the rhino?

The answer is Viagra. For thousands of years, powdered rhinoceros horn has been used as an aphrodisiac in traditional Chinese medicine. Of course, powdered rhino horn doesn't actually work as an aphrodisiac, but that hasn't stopped people from killing huge numbers of rhinoceroses to pursue this deluded belief.

Like much of traditional medicine, the logic behind powdered rhino horn as an aphrodisiac is based on superstition rather than science. A rhinoceros's horn is rigid, curved, and upright, and thus it vaguely resembles an erect penis. Therefore, through a kind of "sympathetic magic" similar to the "doctrine of signatures," rhinoceros horn must be an effective cure for impotence and erectile dysfunction. By a similar logic, tiger penis is another highly prized aphrodisiac in traditional Eastern medicine, because of the tiger's obvious virility and aggression.

As a result, tigers too have been hunted to the brink of extinction.

The growth markets for "traditional" medicine today are not in India and China, but rather the pampered West, where we have enjoyed the fruits of advanced, effective medicine for long enough to take them for granted. In China, people are demanding vaccines, drugs, and effective medicines, all the benefits of an advanced, scientific system that has led to dramatically increased life expectancies in the West. And when it comes to their sex lives, they generally seek out Viagra, Cialis, and other drugs that actually work, rather than powdered rhinoceros horn. Hopefully, this trend, along with the various laws that are designed to protect these animals, will help to stimulate the fertility and population of the rhinos themselves, and allow their numbers to increase.

what is a "paraffinoma"?

We tend to think of cosmetic procedures such as Botox and collagen injections as being recent innovations, the product of today's shallow, vacuous culture, obsessed with youth and beauty. Yet procedures of this type are far from new. As far back as the nineteenth century, doctors were experimenting with similar injections, for exactly the same purposes. In particular, there was a brief trend for paraffin wax injections, which were designed to smooth out wrinkles, just as Botox and collagen do today. Paraffin wax was also injected into breast tissue, in an early attempt at breast augmentation surgery.

There is a very good reason why we no longer hear about paraffin wax injections, as these procedures generally had disastrous results. The injection of paraffin wax causes paraffinomas, which are hard, painful lumps that form into unsightly scars, and can become infected.

Amazingly, despite this, the practice didn't die out completely, and there have been occasional reports of paraffin wax injections throughout the twentieth century. In particular, paraffin wax is sometimes used in attempts to enlarge the penis. One report from 1956 describes a patient suffering from premature ejaculation, who turned to paraffin wax after other treatments had failed. The procedure was not successful. Six years after the injection, his premature ejaculation had not been cured, and he now suffered considerable pain whenever he got an erection. Eventually the patient had the paraffinoma removed surgically, along with the skin of his penis, which was replaced with skin taken from his scrotum. This surgery was successful, although it left him with a smaller scrotum than before.

should babies sleep on their front or their back?

One of the most influential books on childcare was *Baby and Child Care* by the American pediatrician Dr. Benjamin Spock. The book was first published in 1946, and went on to become a record-breaking bestseller, selling more than 50 million copies worldwide and influencing generations of new parents. The book is still widely praised for its sensible and liberal approach, but one of its recommendations has gone dramatically out of fashion.

In the book, Spock recommends that babies should be put to sleep on their front, which is called the "prone" position. His explanation is that if a baby vomits in the night, there is more risk of choking on the vomit if the infant is on its back (which is known as "supine"). Thanks to Spock's book, this became the orthodoxy, and the vast majority of American babies were put to sleep on their front.

However, it now seems that this advice was flawed. Studies into crib death, which is also known as Sudden Infant Death Syndrome or SIDS, have found that sleeping in the prone position increases the risk of crib death, although it's not clear why. There are various theories. A baby sleeping on its front may be more likely to breathe in the same, recycled air, which is higher in carbon dioxide. Alternatively, there may be a greater risk of the baby on its front suffocating itself, or inhaling toxins or mold that some accounts suggest might be present in the mattress.

Although the mechanism is unclear, the statistics are compelling. Since the "Back to Sleep" campaign was launched in 1994 to persuade parents to put their babies to sleep on their back, the rate of supine sleeping in the United States has gone up from 13 percent to 76 percent, while the rate of SIDS has plummeted by more than 50 percent. While this is good news, the alarming implication is that Dr. Spock's well-meaning advice may have led to tens of thousands of unnecessary deaths.

On a lighter note, the *Star Trek* character Mr. Spock was apparently not named after Dr. Benjamin Spock; it is just an odd, perfectly logical coincidence! According to creator Gene Roddenberry, when he was developing his original proposal for a TV show called *Star Trek*, he was just looking for an alien-sounding name, and had never even heard of the pediatrician.

is it true that surgical implements sometimes get left inside patients?

It is true, and it may even happen quite frequently. Doctors and nurses are often overworked and sleep deprived, and they have to perform emergency surgery in a hurry. During a routine operation, an average of 250 to 300 different tools and implements are used, and this number can be as high as 600 in complicated procedures. As a result, it's not entirely surprising that some things occasionally get lost.

A wide range of tools and instruments have been left inside patients, including needles, forceps, knife blades, tweezers, safety pins, scalpels, clamps, towels, and scissors. One patient had to have a nine-and-a-half-inch (24 cm) spatula removed, after it was left inside her following an operation. Another patient recently had a two-foot (61 cm) guide wire removed, which ran all the way from his groin to his upper chest. Apparently, doctors inspected this patient's X-rays six times without noticing the wire. One surgeon even left a set of his own forceps, engraved with his initials, inside a patient.

The most common objects left behind are gauzes, also known as sponges. These are small pads of sterile cotton, which one might imagine would be relatively harmless. However, they can in fact cause considerable harm. After surgery, gauzes left in the body can be mistaken for abscesses or tumors, lead-ing to risky and unnecessary surgery.

One report estimates that there are around 2,700 incidents like this each year in the United States, but this number may well be an underestimate, as many patients are not aware of what has happened, and hospitals are not keen to encourage expensive lawsuits. Patients undergoing emergency surgery face a greater risk of an item being lost, as do obese patients. According to research, a patient with a one-point higher body mass index faces a 10 per-cent higher risk of having some implement left inside them.

Hospitals are trying a number of techniques and new technologies to address this problem, but none of them seems to be foolproof. It is gener-ally standard procedure to conduct an inventory of every tool and implement a number of times at different stages of an operation, and some hospi-tals recommend as many as four separate counts. But there are lots of occasions when there is sim-ply no time for this, and corners have to be cut. It's also not clear how much the counts even help. According to studies, even when all the counts do take place, mistakes still get made. In many cases where an implement has later been found inside the patient, the count that took place after the sur-gery found nothing missing.

There are also technological solutions being developed. One of them involves marking every gauze with "radio-opaque" marking, which will then show up on standard radiograph scans. Another potential solution that has been suggested involves marking every single gauze with an individual bar-code, which is then scanned into a reader at the end of the surgery.

which doctor drank a glass of cholera, and survived?

Decades after John Snow's triumph with the Broad Street water pump, there was still considerable debate about what caused cholera, and how it was transmitted. By 1892, the German doctor Robert Koch had isolated the bacterium responsible for cholera, and was advocating a "germ theory" of disease, which proposed that many diseases were caused by microorganisms, rather than miasmas, bad air, or imbalances of humors in the body.

Koch had not yet done enough to convince the world, however, and there were many rival theories. Max von Pettenkofer preferred the "ground water theory," which proposed that rotting organic material in the soil was what released cholera into the air, and once released, it would only infect those who were particularly susceptible as a result of their poor diet or weak constitution. In other words, the cholera bacterium was not the important factor—the important, decisive factors were the atmosphere and the intrinsic health of the patient. Therefore, von Pettenkofer believed that cholera could not be transmitted from person to person, but only via the atmosphere, and only to those people who were susceptible.

So confident was von Pettenkofer about his assertions that, in front of a number of witnesses, he lifted a beaker of water that contained a growth of Koch's cholera germs, and drank it down. Knowing what we do today, this seems like an absurd risk, something close to suicide. But von Pettenkofer suffered barely any ill effects, apart from a bout of diarrhea, and consequently he felt that in this one balls-out act, he had conclusively disproved Koch's germ theory. The rest of the world largely concurred, and so this one foolhardy act set back the germ theory for decades.

However, we now know that von Pettenkofer was wrong. Cholera is spread by the bacterium, and not by miasmas. So how did he survive? There are a number of possibilities. First, von Pettenkofer may have had high levels of stomach acid, which can protect against cholera when ingested in this way. Second, the cholera sample may have died, perhaps through the presence of an antibacterial mold or bacteriophage. Third, von Pettenkofer may have been immune to cholera; perhaps he had had a very mild bout when he was younger, which he may not have even been aware of, or he may simply have been resistant.

The great Russian composer Peter Ilich Tchaikovsky did something rather similar, but sadly he
was not so lucky. Days after the premiere of his emo-
tional Sixth Symphony, the *Pathétique*, he drank a
glass of cold, unboiled water in a public restaurant,
which was an unthinkable thing to do during a cholera epidemic, and consequently died. The dangers
of cholera and unboiled water were known to everyone, and thus Tchaikovsky's dramatic, public act
has been interpreted by many as a deliberate act of
suicide.

how accurate is the presentation of medicine on tv and in films?

Unsurprisingly perhaps,
the presentation of medi-
cine, hospitals, and surgery
in films and on television
is full of inaccuracies. And
yet, viewers do tend to be-
lieve what they see on the screen. A study in the year
2000 found that more than 50 percent of TV viewers
trusted the medical information on TV, and about
25 percent of people described TV as one of their
top three sources of medical information. Here are
just a few examples of why this is a mistake:

Chloroform—In any kind of TV drama, we
all implicitly understand that one whiff of a

chloroform-laced handkerchief will render any person silently and harmlessly unconscious within a second or two. In reality, the effects of chloroform are not nearly so predictable. Chloroform often takes some time to act, perhaps as long as a minute, and in that time the victim would be likely to struggle, shout, fight back, and generally make things difficult. On the other hand, it's also very easy to give too high a dose of chloroform, which can cause vomiting, brain damage, and even death. One reason that chloroform is hardly ever used as an anesthetic these days is that the margin between adequate anesthesia and a potentially fatal dose is far too fine a line.

Trauma—In TV and films, it's perfectly routine for characters to get whacked with bottles, chairs, guns, and a whole range of other weapons, without suffering any real consequences. People simply fall over, and then get up a bit later, and they're fine. In reality, just one of these blows to the head can cause concussion, brain damage, memory loss, and a loss of vision or hearing. When the victim regains consciousness, if they haven't suffered a blood clot, or slipped into a coma, they will certainly be extremely groggy and nauseous, and in no fit state to leap into a car and chase the villain into the next scene.

CPR—Cardiopulmonary resuscitation, or CPR, is that thing they do in hospital dramas when they pump the patient's chest with their palms, and give them mouth-to-mouth (or, increasingly, ventilate the lungs using a bag or some other device). A recent study found that in TV dramas, more than two-thirds of CPR patients survived. In reality, the number is more like 15 percent.

Heart attacks—This is another example of TV shorthand that we all instinctively understand: The fat man clutches his heart, gasps, and keels over and dies, and every viewer knows that this is the code for "heart attack." But heart attacks rarely look like this in real life. They don't tend to cause instant death, and often there is no sharp, sudden pain in the heart. This misinformation has real, harmful consequences, as people are often slow to call an ambulance when they do have a heart attack, because they don't recognize it as a heart attack. The real symptoms of a heart attack are often much milder than TV suggests. They include nausea, shortness of breath, sweating, dizziness, numbness in the arms, and pain in the chest, neck, or shoulders, but not the sharp, stabbing kind. Women in particular almost never experience the sharp chest pain so beloved of TV drama.

Defibrillators—In any hospital drama, no matter how bad the patient is, there is always a last resort. If he appears to have died, and the monitor has flatlined, you can always just shock him back to life, with the two magic electrical paddles, the defibrillators. In reality, paddles are rarely used these days; sticky pads are applied instead. More important, defibrillators simply don't do the thing that TV producers all seem to think they do. They don't shock the heart back into life. All they can do is correct a heart's rhythm. There are also no sparks in real life, and the patient's body doesn't leap into the air in a dramatic convulsion.

Guns—Most action heroes would be deaf by the third reel. Gunfire is extremely loud, and frequent exposure to it can cause permanent hearing loss. Handguns and automatic weapons fire at around 140 decibels, which is defined as the threshold of physical pain, and can cause instant hearing loss. Larger weapons such as bazookas and rocket-launchers are even louder. Guns can also damage your eyes. Shooting ranges provide eye protection from the threat of spent shells and gunpowder dust, but this is never an issue on TV.

People also never fire guns properly. The recoil from even a small handgun can knock a man over if he's not properly braced, firing straight from the shoulder. In real life, police officers don't shoot one-handed, or side on (in fact, most police officers never fire a gun once in the line of duty, throughout their entire careers). According to reports, a common injury among gang members is a broken thumb, caused by recoil from mimicking the Hollywood trend for sideways firing. As for firing automatic weapons with one arm, it's not advisable, as it will very likely dislocate your shoulder.

Finally, being hit by a bullet does not send the

victim flying backward. Bullets are small, they don't weigh very much, and much of their energy is spent in traveling through your body, rather than pushing it backward. In fact, many people who have been shot report that they didn't feel a thing when the bullet hit them; the pain and shock tend to come later, when you actually see the wound and the blood.

what was surprising about the british government's press conference that finally announced a proven link between smoking and lung cancer?

In the first half of the twentieth century, the number of deaths from lung cancer began to grow rapidly. As the prevalence of contagious diseases such as tuberculosis and smallpox decreased, these deaths were being replaced by noncontagious causes of death: lung cancer, heart attacks, and strokes. Between 1905 and 1945, the rate of lung cancer in men increased twentyfold. In 1950, the number of deaths from lung cancer exceeded those from tuberculosis for the first time.

No one knew the cause of this increase. There was no evidence that smoking was harmful for adults. There were restrictions on cigarettes being sold to children, but only because it was believed to stunt their growth. In 1950, Richard Doll and Austin Bradford Hill published the first major study to demonstrate that lung cancer was caused by smoking. As a result, Doll himself gave up smoking.

The 1950 study was a case-control study, which compared one group of 709 people who had lung cancer with another group of 709 who were as similar to them as possible in every respect, except that they did not have lung cancer. The aim of a case control study such as this is to see if, all other things being equal, one factor leaps out as being distinctive to one group. In this case, it did. Of the men who had lung cancer, only two were non-smokers, whereas twenty-seven of the men without lung cancer didn't smoke. This was a statistically significant finding that would be confirmed by further studies in 1954 and 1957 by the same authors.

By 1954, the case was undeniable: Smoking tobacco enormously increased the risk of lung cancer. The British government's minister of health, Iain Macleod, held a press conference in which he announced that lung cancer was caused by smoking. There was one extraordinary detail about this press conference—Macleod chain-smoked throughout it.

why is it that so many murderers are doctors?

There do seem to be a dispropor-
tionate number of medical professionals among history's ranks of
murderers, but it's not clear why
this should be. It's possible that
doctor-murderers are more memorable and newsworthy; when a
doctor kills it is particularly shocking and frighten-
ing, because we invest so much trust in our doctors. Doctors are in a unique position to cause harm
if they so choose, and cover their tracks, and most
people's knowledge of medicine is so rudimentary
that the balance of power and knowledge is entirely
in the doctor's favor.

Alternatively, it could be that doctors kill more
often simply because it's easier for them to do so.
Doctors have access to a wide range of poisons, and
an expert knowledge of how they can be used. A
doctor will know better than most which poisons
can be hidden in food or drink, how long before
they take effect, what the symptoms might be confused with, and so on. A doctor is also in a unique
position to administer such a poison, and less likely
to be squeamish about the outcome. Many doctors
also have access to surgical tools, and other lethal
weapons, as well as a particular knowledge of the

human body, where it is weak, and how it can be damaged, numbed, cut open, and so on. But do any of these explanations account for the actions of these men?

First, of course, comes Dr. Harold Shipman, one of the most prolific serial killers in history. He is known to have been responsible for an astonishing 215 murders, but the actual number is likely to be much higher. Most of Shipman's victims were elderly female patients in the UK. He would give them lethal doses of diamorphine, sign their death certificates himself, and then forge their medical records to give the impression that they had been ill. In the year 2000, Shipman was sentenced to fifteen consecutive life sentences, and the Home Secretary ordered that he never be released. In 2004, he committed suicide, hanging himself in his cell. Amazingly, given the stories that follow, Shipman is the only doctor in British legal history to have been found guilty of murdering his patients.

Dr. Edme Castaing was executed in France in 1824. He killed a wealthy patient named Hippolyte Ballet and his brother Auguste using morphine, making him the first person to use the new drug to commit murder.

British General Practitioner John Bodkin Adams was suspected of killing more than 160 of his patients in the 1950s, most of whom died in "suspicious circumstances," and left him money in their wills. In 1957, Adams was charged with murdering wealthy patient Edith Morrell with a lethal dose of barbiturates, heroin, and morphine. Even though he arranged for her cremation on the same day as her death, he was somehow acquitted.

Dr. Buck Ruxton of Blackpool, England, in 1935 cut up the bodies of his wife and maid, covering the carpets and curtains with blood. When he was quizzed by the police, he said he'd cut his hand on a tin of peaches.

Dr. Henry H. Holmes was America's first serial killer. He opened a hotel in Chicago for the 1893 World's Fair, which he built and designed himself. The upper two floors were a maze of dead ends, windowless rooms, stairways to nowhere, and other features designed to confuse the unwary. Here, he would lure numerous women, including employees and lovers, and torture and kill them. Some rooms had gas lines, through which he would asphyxiate his victims. The victims' bodies went down a chute to the basement, where Holmes would dissect them, or sell them to medical schools. In 1896, Holmes was hanged at Philadelphia County Prison. Over the course of his life, he is believed to have killed as many as one hundred people in cities all over the United States.

In the 1930s, Dr. Morris Bolber of Philadelphia carried out an audacious scheme, murdering thirty patients with poison, and sandbags to the head, so that he could collect their insurance money.

Dr. Hawley Harvey Crippen murdered his wife

with an overdose of hyoscine and buried her in the cellar (apart from her head, which is still missing). He then fled to Canada with his lover on an ocean liner. During the voyage, she disguised herself as a boy throughout. However, Crippen was recognized on the ship, and a series of Morse code communications led to him being arrested on his arrival in Canada.

Finally, Dr. William Palmer of Rugeley, England, had a reputation for extravagant living, and an eye for the ladies. He was also an inveterate gambler, but not a successful one. After one of Palmer's horse racing friends, John Parsons Cook, won a large sum of money at Shrewsbury, Palmer invited Cook to dinner to celebrate. After dinner, Cook became violently ill, and died two days later. At the postmortem, Palmer tried to pocket Cook's stomach. He then tried to bribe various people involved with the coroner's office. Eventually, Palmer was charged with murder, after it was found that he had bought strychnine days before the murder.

chapter seven

the human body

The art of medicine consists in amusing
the patient while nature cures the disease.

VOLTAIRE

what are hiccups for?

When a person hiccups, the diaphragm pulls down abruptly, drawing air into the lungs. Within 35 milliseconds, the opening at the top of the air passage, which is called the glottis, slams shut, causing that strange "hic" sound. Most bouts of hiccups end quickly, but some unfortunate people suffer for days and weeks on end. The world record is held by Iowan Charles Osborne, who hiccupped continuously for an astonishing sixty-nine years, until his death in 1991 (which was not caused by hiccups). Despite his condition, Osborne managed to lead a relatively normal life, even somehow marrying, and conceiving eight children.

So what causes hiccups? There are many theories, and there seem to be many different causes. Some people say that hiccups are caused by laughing, gulping, drinking alcoholic or fizzy drinks, or swallowing too much air. Some blame spicy foods, or rich foods, or overeating. There are medical conditions that can cause hiccupping, including epilepsy, diabetes, tuberculosis, and bowel obstruction. One report suggests a man hiccupped for four days straight, because a hair inside his ear was tickling the eardrum.

One persuasive theory holds that hiccups can be caused by the vagus and phrenic nerve systems. Persistent hiccups are often caused by nerve damage; chemotherapy can cause hiccups, as can tumors found at the site of these nerves. One persistent hiccup sufferer was found to have a brain tumor; once two-thirds of the tumor had been removed, the hiccups stopped.

An interesting theory has recently emerged that attempts to explain why we have this useless, annoying reflex. Christian Strauss from the Pitié-Salpêtrière Hospital in Paris argued that hiccups are a vestigial evolutionary relic, left over from a very distant ancestor who had gills like a tadpole. Hiccupping is similar to the way primitive air-breathing creatures like tadpoles gulp in air, while closing their glottis to prevent water getting into the lungs. Ultrasound scans have revealed that babies start hiccupping in the womb after just twelve weeks, before any breathing functions have developed. Perhaps hiccupping is an important stage in the baby's growth, preventing amniotic fluid from entering the lungs. If so, it would explain why humans still retain the ability to hiccup, even though as adults it serves no useful purpose.

yes, yes, yes . . . but how do you cure them?

I'm afraid there's even less consensus on this point. Everyone seems to have their own pet theory, but there is no strong evidence in favor of any one. Some of the more common suggestions include drinking water (often in some bizarre contorted position), drinking vinegar, blowing up a paper bag, receiving a fright, controlled breathing, sneezing, sucking, holding your breath, pulling your tongue, pressing your eyeballs, or eating a spoonful of sugar.

If the bout goes on for a dramatic length of time, there are also medical interventions that can be tried. Options include hypnosis, acupuncture, a catheter through the nose, or drugs, including sedatives and anti-spasmodics. These are only for seriously persistent cases, however, as effective treatment with sedatives usually requires a dose that will knock the patient unconscious.

how long does it take to digest chewing gum?

I'm sure we all remember being warned as children not to swallow our chewing gum because, as we all know, it stays in the stomach for seven years. In fact, it's simply a myth, which has presumably emerged from our observations that a) chewing gum isn't food, and so shouldn't be swallowed, and b) chewing gum seems to have the strange property of being indestructible. Therefore, it is indestructible and should not be swallowed.

In fact, chewing gum does not remain in the stomach for any longer than any other foodstuff. The functions of our digestive system, broadly speaking, are to digest those things we can use, and expel the rest. There are many things that the stomach struggles to break down, such as sweet corn kernels, but it's not a problem; these simply get expelled in our poo. It's true that chewing gum is largely indigestible, so it gets expelled in the same way.

Chewing gum generally consists of gum base, flavorings, sweeteners, and softeners. The last three can all be digested, which is why gum quickly loses its flavor, sweetness, and softness as we chew it. Gum base, though, is indigestible. Today, gum base is usually made of butyl rubber, which is also used in the manufacture of inner tubes. In the early twentieth century, it was usually made of natural resin from the chicle tree, which is native to Mexico. Mexicans have chewed chicle resin for centuries.

In fact, people have chewed all sorts of gums for thousands of years. Lumps of tree tar with tooth prints in them have been found in Finland dating back five thousand years; they are believed to have been used to improve oral hygiene, and the tar may also have had antiseptic properties. The ancient Greeks chewed gum, as did Native Americans. Today, the U.S. military provides chewing gum to troops, as it helps to improve concentration, and aids oral hygiene. The army has also begun providing chewing gum laced with caffeine to keep soldiers alert in the field.

Although chewing gum generally doesn't get stuck in the stomach, it is possible to create a blockage if you swallow enough of it, quickly enough, in combination with other indigestible items. A recent report in *Scientific American* cited three cases of children who had swallowed lots of gum, along with coins, or sunflower seeds, and had consequently managed to create gooey, lumpy masses that did get stuck in their intestines, requiring medical treatment. So the answer is clear: Chewing gum doesn't get stuck in the stomach. Except when it does.

what are the crypts of lieberkühn?

It sounds like the title for a new Indiana Jones movie, but in fact the crypts of Lieberkühn are a part of the human body. They are glands found in the lining of the small intestine and colon. In the context of anatomy, a crypt means a pit or dead-end tube in an otherwise flat surface. The crypts of Lieberkühn produce enzymes, including sucrase and maltase, which break down food. The crypts were named after the eighteenth-century German anatomist Johann Nathanael Lieberkühn, who is now best remembered for the medical specimens he produced by injecting wax into various organs and bodily cavities.

In fact, many of the great pioneers of medicine live on through the obscure parts of the body that bear their name. The following are, in my humble view, some of the more exotic and amusing examples:

- The fissure of Rolando is a fold in the cerebral cortex of the brain, named after Italian anatomist Luigi Rolando.
- The foramen of Winslow is a connection between two cavities of the abdomen, named after Danish anatomist Jacob Winslow.
- Adamson's fringe is a part of the hair follicle, where the active hair bulb meets the inactive hair shaft. It was discovered by dermatologist Horatio George Adamson.
- The ampulla of Vater is formed where the common bile duct meets the pancreatic duct. It was first described by a German anatomist named Abraham Vater.
- The sheath of Schwann is a microscopic membrane surrounding one of the cell

types that make up a nerve fiber. It is named after German physiologist Theodor Schwann, whose many accomplishments include discovering the organic nature of yeast.

- The circle of Willis is a circle of arteries that supply blood to the brain, named after Thomas Willis, a pioneering researcher into the anatomy of the brain.
- Scarpa's triangle is a region on the upper thigh, where a number of major arteries and veins are found directly under the skin, making it an important site for surgery. It was named after Antonio Scarpa, a leading Italian anatomist, whose head was removed and exhibited at the Institute of Anatomy after his death.

do people go mad when there's a full moon?

 For thousands of years, people have known that the moon's cycle affects our behavior. The words *lunacy* and *lunatic* come from the name of the

Roman goddess of the moon, Luna. In the seventeenth century, England's chief justice Sir William
Hale was in no doubt about the moon's power
when he stated, "The moon has great influence on
all diseases of the brain, especially dementia."

And the evidence continues to pile up. We have
all heard anecdotal evidence: how emergency rooms
are always busier on a full moon, schoolchildren
become even more unruly, and violent crime goes
up. But there are also serious studies. In the year
2000, a Finnish research project found that people
were more likely to commit suicide when there was
a new moon. In 2007, a police force in Brighton,
England, studied their statistics and found evidence
of a direct link between the full moon and increased
violence on the streets.

And yet, despite all this evidence, it is surely a
myth. There is no meaningful correlation between
the cycles of the moon and any of the reported
human behavior. How, though, you may be wondering, could anyone argue with the compelling
evidence already cited?

Well, there are a number of factors that can
explain this discrepancy. First, there is the possibility of random variation: If you look at any set of
statistics, some kind of pattern is likely to emerge.
The question is whether or not this pattern is significant, and repeated. Although there are a few
studies that do appear to show behavior changing
during specific parts of the moon's cycle, there are

also many others showing no effect, or even showing contradictory results. For example, two studies published in the December 23, 2000, issue of
the *British Medical Journal* looked at the rate of dog
bites around the full moon: One study found that
rates increased, the other study found that they
decreased. Taken together, they provide no evidence of any effect. In 1986, scientists at the University of Saskatchewan carried out a meta-analysis,
combining thirty-seven separate studies that had
looked at the moon's effect on various aspects of
human behavior. Overall, the scientists concluded
many of the studies' reported findings were based
on small samples, contradicted one another, and
that there was no evidence of any lunar effect.

Another problem is simple human error. Statistics are complicated, and many studies are
flawed. The same meta-analysis looked at twenty-
three studies that had claimed to find a correlation
between human craziness and the moon's cycle, but
found that nearly half of those studies contained at
least one statistical error. Other meta-analyses have
since confirmed these findings: There is no overall
evidence of correlation, and those studies that do
appear to show a correlation are often methodolog-
ically flawed.

This problem is made worse by the media, which
only ever reports stories that confirm the crazy
moon theory. This is to some extent understand-
able. When a study finds that there is, say, no link

between homicide rates and full moons, that's not much of a story. When, however, a study comes out a week later with the opposite finding, that makes a great story, it doesn't really do any harm, and besides, editors have got pages to fill.

Another factor is confirmation bias, which essentially means that people tend to notice things that fit their existing beliefs or hypotheses. Therefore, if you work in an ER, for example, and you've heard that it gets busy when there's a full moon, you may well remember those nights when it does happen to be busy and there's a full moon. Perhaps you have a coffee with your colleagues when it's over, and swap stories of other crazy full moon nights. Busy nights when there's no full moon, or quiet nights when there is one, may well be less memorable.

Finally, it's not clear how a full moon could have any meaningful effect on human behavior. Those who believe in the "lunar effect" tend to mention the tides. They argue that 80 percent of the earth's surface is water, just as 80 percent of our body is made up of water, and that therefore we might be affected by the moon's gravity. But there are a number of problems with this. First, a full moon is just more visible than other moons, it doesn't have more gravity. A new moon is just as close to the earth as a full moon, so it ought to have the same gravitational effect. The second problem is that the moon's effect on the tides is tiny; it is estimated that a mother holding her baby is exerting around 12 million times as much gravitational force on the child as the moon, simply because she is closer to it. One astronomer has estimated that a mosquito landing on your arm exerts more gravitational force on you than the moon.

ah, but we already know that the moon can affect people, because we know it affects the menstrual cycle!?

Well actually, the evidence suggests that it doesn't. The apparent reason for the development of this myth is that the lunar cycle, which is the length of time between two full moons, is about 29.53 days, which is similar to the average length of the menstrual cycle. However, although the average is about 28 to 29 days, the lengths of women's cycles vary enormously: Women can have regular cycles as short as 21 days, or as long as 36. Longer or shorter cycles will, by definition, be out of sync with a 29.53-day lunar cycle. However, cycles of around 26 to 30 days might appear to be in sync just often enough to reinforce this myth.

The cycles may also appear to be synchronized because there are a number of different phases of the menstrual cycle, and a number of different phases of the moon, many of which generally last for a few days rather than a single day, making it easier to link any two of them together. For example, one study in 1980 found that 40 percent of women in a random sample began their periods during the "light half cycle," meaning the two weeks around the full moon. But two weeks is a very broad period of time. If women's cycles were not synchronized with the moon, we would expect around 40 percent of them to begin in a given two-week period. In other words, 40 percent is exactly what we'd expect if the moon had no effect on women's cycles.

There is also an established belief that women synchronize their menstrual cycles with one another, but this too seems to be false.

The menstrual synchrony theory received a major boost in 1971 when Martha McClintock published a paper in *Nature* magazine which found that women in her dorm at Wellesley College, Massachusetts tended to synchronize their periods. Yet the study only suggested that the women's cycles were, on average, within about five days of one another. This may sound compelling at first, but it's pretty much what we would expect from random variation in such a small sample. If the average cycle is 28 days, then two women's cycles could not be more than 14 days apart; so this is the maximum difference. The minimum difference is zero. We would therefore expect an average of about 7 days' difference, with just as many incidences of cycles being fewer than 7 days apart as there are incidences where the gap is greater than 7 days. In this context, in a small sample, an average of 5 days is not significant.

There were also a number of methodological problems with McClintock's study. First, the women reported their dates themselves, retrospectively, which creates the possibility of confirmation bias. Second, which cluster of dates do you decide to use? In any small sample of random data such as this, it's fairly easy to pick a point where there is an apparent cluster. Further studies have attempted to replicate these results, but have found no evidence of menstrual synchrony.

where in the human body is the soul found?

The French philosopher René Descartes was fascinated by a tiny gland called the pineal gland, which is found deep in the center of the brain. This tiny teardrop-shaped gland seemed to serve no purpose, and appeared to be the only part of the brain that consisted of one single unit, rather than two matching pairs or halves. Since our senses generally work in pairs—we have two eyes, two ears, and so on—Descartes's theory was that there must be

a central hub in the brain where information col-
lected in this dual fashion was brought together
and assimilated, to be presented as a single sensory
experience to that part of us that we call the mind
or the soul. Therefore, Descartes concluded, the
pineal gland must be where our soul is located, and
where all our thoughts and ideas are formed.

Today we know a little more about how the
pineal gland works, along with the rest of the brain.
The pineal gland produces a hormone called mela-
tonin, which regulates puberty. It also regulates
our circadian rhythm, which is our body's biological clock. It determines which part of the day we
feel most awake, most drowsy, most coordinated,
and so on.

However, people of a mystical or spiritual bent
also have a number of more exotic theories about
this mysterious part of the body. It is sometimes
described as our "third eye," because it seems to
have some evolutionary link with the photorecep-
tive parietal (or third) eye that is found in some
other animal species, including lizards, frogs, and
lampreys. There are claims that the pineal gland
can secrete a natural psychedelic drug called DMT,
which is linked to near-death experiences such as
out-of-body experiences, visions of a light at the end
of a tunnel, encounters with angels, and so on. The
pineal gland also seems to affect our responses to
certain drugs, such as cocaine and antidepressants.

The pineal gland does not seem to be the site of the soul, but it remains mysterious and poorly understood. In 2007, an experiment in China using MRI scans found that the area surrounding the pineal gland becomes more active during a religious meditation technique called Chinese original quiet sitting, in which practitioners silently recite religious mantras. Although this was a very small-scale study, it has added to the gland's mystique as our mysterious, powerful connection to the spirit world.

do london taxi drivers have bigger brains?

To drive a London black cab, taxi drivers have to pass an extremely demanding course called "The Knowledge," which requires them to develop an intimate familiarity with every street and place of interest in the city. It can take three years to learn, and more than three-quarters of those who start the course drop out.

In the year 2000, a study by Eleanor Maguire at University College London found that London cabbies tend to have one region of the brain that is significantly enlarged, compared with the rest of us. At this point, you might be wondering what this means. Does it indicate that only brainy people are capable of learning "The Knowledge"? Or alternatively, does it suggest that driving a taxi enlarges the brain? Those are very sensible questions. In fact, the study in question does attempt to distinguish between these two possibilities. The research team found that the size of the posterior hippocampus was progressively bigger in taxi drivers depending upon how long they had driven a taxi, which indicates that driving a cab does make your brain expand.

Yet the researchers also found that there was a trade-off. While taxi drivers did develop a larger posterior hippocampus, which suggests that this part of the brain is used in navigation, their anterior hippocampi would shrink by an equivalent amount to compensate, so overall their brains were no bigger than average. This may even mean that taxi drivers lose some of their other faculties in the course of developing their navigational expertise.

what percentage of our brain do we use?

It is a well-known fact that we only use 10 percent of our brain, and that therefore we have enormous capacity to improve our intellectual abilities, and harness mysterious powers. Sadly, though, this is another of those well-known facts that turns out to be false. The very specific nature of this claim— it is almost always 10 percent, rather than, say, "a small proportion"—adds to the impression that it is based on a particular scientific study, but in fact there is no study, and this fact seems to simply be a misconception that somehow took hold.

The idea that we only use 10 percent of our brain seems to have developed during the late nineteenth century. It has been attributed to the pioneering psychologist William James, who, in his book *The Energies of Men*, stated, "We are making use of only a small part of our possible mental and physical resources." The anthropologist Margaret Mead is believed to have said something similar, and Albert Einstein is said to have made some similar quip concerning his own intelligence. One way or another, the idea took hold that this was a proven, scientific fact.

Still, it's hard to see how it could actually be true in any meaningful way. If 90 percent of our brain was really not being used, then people could lose 90 percent of their brain in an accident, and suffer no ill effects. In reality, a small amount of damage to any area of the brain will usually reduce a person's abilities to a dramatic degree. If 90 percent of our brain were removed, we would be left with about 140 grams of brain, which is about as much as a sheep has. Sheep are cute, but they're not very clever.

Another objection is that most parts of our body tend to wither if they are not used. When people break a leg, and spend weeks in a plaster cast, the limb becomes noticeably weakened through under-use. If 90 percent of our brain was never used, we would expect adults to have shrunken, withered brains, but they don't.

Furthermore, if 90 percent of our brain were never used, it seems reasonably likely that we would probably evolve to have smaller, less demanding brains, perhaps even smaller skulls, to make childbirth slightly easier. Our brains comprise only around 3 percent of our body weight, but they use up an enormous 20 percent of our energy and resources. If 90 percent of the brain were made of useless material, which was nevertheless costing us all this energy, isn't it likely that we might have evolved a more efficient system?

Although the 10 percent "fact" is false, it's possible to imagine how the misconception may have arisen, or been reinforced. One explanation is that only 10 percent of our brain consists of neurons, the other 90 percent is made up of glial cells, which encapsulate and support the neurons. In the past, glial cells were thought to have only a very limited function, but recent studies have shown that, like neurons, glial cells have chemical synapses, and release neurotransmitters. People may have interpreted this ratio and drawn the conclusion that we only use our neurons, which make up 10 percent of the brain's cells. Alternatively, the misconception may have arisen in reference to the number of neurons that are firing at any one time, which may well be around 10 percent, although it's not immediately obvious how this could be measured.

However it arose, we now know that the theory is false. Brain imaging scans of various types have shown that we use every part of our brain, and most sections of the brain are active at any one time, even while we are asleep.

are women's breasts getting bigger?

They are indeed. In 1983, the best-selling bra sizes in the United States were 34B and 36B; today the bestseller is 36DD, which is a significant increase. In the UK, Marks & Spencer bras used to go up to a G size, but in 2007, the company announced that it was introducing sizes GG, H, HH, and J, to accommodate the larger busts of modern women. So what's going on here?

The obvious answer is that we are simply getting bigger in general. Since 2007, the rate of obesity in American women has grown by 2.1 percent to 35.3 percent. Women are better nourished, which means they are taller, with a stronger frame. The growth in breast implants may also be a factor, although probably not a significant one. In 2007, just over 300,000 breast augmentation procedures took place in the United States, which may sound like a lot, but isn't really a huge number in a population of 155 million women. Another factor is the increasing use of the contraceptive pill, as breast size is to a large degree determined by hormones.

Some people also credit improved education about properly fitting bras. According to various surveys, anywhere between 70 and 90 percent of women wear the wrong size bra, and some experts suggest that the very system of bra measurement is so flawed as to be practically useless. In November 2005, Oprah Winfrey devoted an entire show to bras and finding the correct fit, and this TV event encouraged vast numbers of women to have a proper fitting carried out. In many cases, this meant changing to a smaller band size, but a larger cup size.

Intriguingly, there is also something called "cup inflation" going on. The reverse trend is well known in fashion retailing: Increasingly, dress sizes are getting smaller, which means that larger and larger women can squeeze into what is labeled as a size 4 and feel good about themselves. This is one reason for the enduring fascination with Marilyn Monroe's dress size. It is widely claimed that Monroe was a size 16, which has led to a lot of debate about whether models today are too skinny, and so on, but in fact, even if Monroe was a size 16 in the 1950s, that doesn't mean she would be a size 16 today. Recently, a British fashion writer was given the chance to try on some of Monroe's dresses, and found that they were equivalent to a UK size 8 or 10 today, which is roughly a 4 or 6 in the US.

Cup inflation, on the other hand, is moving in the other direction. Increasingly, manufacturers are labeling bra sizes bigger than in the past, so that customers will feel reassuringly buxom and sexy. A woman who was a 36C a few years ago might well now be a 36DD, without any change in the actual size of her breasts.

There is one last interesting detail. It's not only women's breasts that are getting bigger. Men's are too. The problem of man-boobs, or gynecomastia to give it its proper name, is a growing one. It is thought to be caused by hormone imbalances, and by excess body fat. Weight loss can help the problem, but many patients also consider surgery.

is the tongue the strongest muscle in the body?

It's not clear how this idea has gained currency, as it's hard to imagine any definition of strength by which the tongue would be considered the strongest muscle in the body. The most sensible definition of a hypothetical "strongest muscle" would seem to be the one that can apply the greatest direct, measurable force on some external object, in which case the winner is the masseter muscle. The masseter is the jaw muscle, and there are two of them, on either side of the jaw. The masseter has a significant advantage over other muscles in that it uses the jawbones as powerful levers. As a result, jaw muscles have been recorded delivering a bite force of 975 pounds (442 kg) for 2 seconds, according to the *Guinness Book of World Records*.

Alternatively, if we consider muscles without the mechanical advantage of levers of this type, then the strongest muscles are simply the biggest ones, as individual muscle fibers are generally all of similar strength. By this measure, the biggest muscles are either the quadriceps, which are the front thigh muscles, or the gluteus maximus, which is basically the bum.

There are other ways we might define strength too. Pound for pound, shorter muscles tend to be stronger than longer ones. Relative to its weight, the myometrial layer of the uterus is the strongest muscle in the body, as the entire uterus only weighs about 2.2 pounds (1 kg), but exerts a force during childbirth of as much as 100 pounds (a little more than 45 kg). To look at the question in another way, the heart does more work over the course of a lifetime than any other muscle. Most muscles quickly get tired when given work to do, but the heart beats continuously. In short, there are any number of different ways to work out what we might consider to be the strongest muscle in the body, but the tongue doesn't seem to be the answer no matter which method is used.

chapter eight

rude bits

Always laugh when you can. It is cheap medicine.

LORD BYRON

did christopher columbus bring syphilis to europe from the new world?

This question continues to be the subject of heated and enduring debate. The theory goes that when Christopher Columbus returned from his first trip to the Americas in 1493, some members of his expedition brought back syphilis, which they had contracted in the New World. Within two years, Europe had its first epidemic of syphilis, which was to plague the continent for another four hundred years.

This is by no means a new theory. For centuries, people have believed that Columbus and his expedition brought syphilis to Europe, and there is considerable evidence to support the idea. The first European syphilis epidemic broke out in 1495 among French troops, who are believed to have included in their number many of Columbus's mercenaries. There is also archaeological evidence, as the bones of syphilis victims have distinctive lesions and pitted skulls. Lots of New World, pre-Columbian remains have been found with these markings, which suggests that syphilis was certainly present in South America before Columbus's arrival. A recent study looked at the molecular structure of twenty-six strains of the bacterium *Treponema pallidum*, and found that those strains that cause venereal syphilis were the ones which had originated most recently, and were closely related to similar strains from South America.

The theory also has its critics, as there is some evidence that syphilis was already present in Europe before Columbus's return, but cases of it may often have been described as leprosy. As far back as the fourth century B.C., Hippocrates described a disease that some believe may have been syphilis. Fourteenth-century skeleton fragments from England and Italy have been found that possibly show signs of syphilis, although this is disputed. In response to the recent molecular study mentioned above, a number of critical anthropologists have argued that the strains of *Treponema pallidum* exhibit fewer similarities than the study claims, and that as a result no firm conclusions can be drawn.

More generally, the idea of a disease being specifically blamed on one person, expedition, or nation is something many people find uncomfortable, and potentially loaded with colonial implications. After all, if Columbus brought syphilis from the New World, does that mean that syphilis came from the Americas, that it was somehow the fault or responsibility of the invaded people? Alternatively, is it reasonable to pin the blame on Columbus for being an unwitting vector for an unknown disease, in an age before we even knew germs existed?

The issue is further complicated by the fact that

the diseases Columbus took with him from Europe to the New World had a far more devastating effect than syphilis. The New World natives had no resistance or immunity to the germs introduced by the conquistadores. The first epidemic may have been swine flu, transmitted by pigs on Columbus's ships. This was followed by smallpox epidemics on Hispaniola, Puerto Rico, and Cuba. When the Spanish adventurer Cortés attacked the main Aztec city of Tenochitlán (today's Mexico City), he took just three hundred troops. Three months later, the city had fallen, and more than half of its population of 300,000 had died, most of them from smallpox. More waves of disease followed, as Amerindian populations were decimated by measles, typhus, smallpox, and influenza. Many native populations were brought to the brink of extinction. As a result, Columbus's arrival at Hispaniola has been described as "the most disastrous event in the history of human health."

why were obstetrical forceps kept secret for over a century?

Giving birth in the sixteenth century was an extraordinarily painful and dangerous business. If there were any complications, the odds of mother or baby surviving were slim. Toward the end of the century, a pair of French brothers began to make their name in London. They were the sons of Huguenot surgeon William Chamberlen, who had fled France in 1576. Bizarrely, both sons had been named Peter, so they were known as Peter the Elder and Peter the Younger.

The family business was surgery, and obstetrics in particular, even though male midwives were rare at this time. Peter the Elder became the surgeon and obstetrician to Queen Anne of Denmark, in London, and as their reputation grew, the brothers became celebrated and wealthy, with Peter the Elder attending a number of senior royal births.

The reason for their success was that they had a tremendous secret, which they publicly alluded to. They hinted that they had certain techniques, or perhaps instruments, that allowed them to safely deliver babies in circumstances that other mid-wives would find impossible. Of course, if this were true, it could potentially be of enormous public benefit, and save many lives, but there was no hope of patenting such an instrument or technique, and so the Chamberlens felt that it was in their commercial interest to keep it a secret.

To protect their secret, they went to extraordinary lengths. They would bring the device into the birthing room in an enormous ornate box, which took two men to carry. The idea was clearly to give the impression that the box contained some huge, complicated machinery. The expectant mother would be blindfolded, to prevent her from seeing the device, and no observers were allowed in the room. Instead, they would listen on tenterhooks outside the door, while ringing bells and strange mechanical noises hinted at magical technologies.

Of course, this was all smoke and mirrors. The device was simply the obstetrical forceps, which is a curved, two-bladed instrument that is still used today in difficult births. It is believed that Peter the Elder invented the device. The Chamberlen family business continued well into the eighteenth century, as Peter the Younger's son, grandson, and great-grandson all carried on the tradition, and no one managed to find out the secret of the Chamberlen family's remarkable success in delivering difficult births. At various points, they even tried to sell the device, and eventually Peter's grandson Hugh is believed to have done so, selling the forceps in the Netherlands to a select group of doctors. However, the secret didn't come into the public domain until the 1720s, perhaps 120 years or more after its invention.

In 1650, the rate of maternal death in childbirth was around 160 for every 10,000 births. Thanks to the widespread adoption of obstetrical forceps, by 1850 this had dropped to just 55 deaths per 10,000 births. For more than a century, almost three times as many women died in childbirth as was necessary. In 1813, the Chamberlen family forceps were finally found, hidden under a trap door in a loft at the family's grand country home in Essex.

did a large-breasted girl inspire the invention of the stethoscope?

In early nineteenth-century France, doctors would listen to a patient's heartbeat or lungs by simply placing their ear against the patient's naked chest. There are even reports of weary doctors laying their head on a soft, warm bosom and accidentally nodding off to sleep. But some doctors found this method ineffective, as well as somewhat embarrassing, and one of them was the French physician René Laennec.

One day in 1816, Laennec found himself faced with a patient whose generous embonpoint made such an inspection unfeasible. Laennec had an idea, perhaps aided by the fact that he played the flute: "I happened to recall a well known acoustical phenomenon: If one places one's ear at the end of a beam, one can hear very distinctly a pin dropped on to the other end." Thus inspired, he rolled up a few sheets of paper, fashioning a rudimentary stethoscope, through which he could listen to the girl's heartbeat from a safe distance, without having to go near any boobs.

To his surprise, Laennec found that his device was not just adequate, but clearly superior. He could hear the heartbeat far more clearly than he had ever previously managed by laying his ear against the patient's chest. Fired with success, he set about inventing a proper medical tool; his first stethoscope was essentially a long wooden cylinder, which he then went on to refine and develop. Using this device, he was able to observe and classify the different sounds produced by the heart and lungs to a much greater degree of detail and clarity than had ever been done before. Many of the terms used by doctors today to refer to the various sounds produced by the heart and lungs were first recorded and defined by Laennec.

what was unusual about
miss jenny's launderette
in georgian london?

The answer is that it was a condom launderette. At Miss Jenny's, on St. Martin's Lane near Charing Cross, ardent young men could have their condoms, which were made of linen or animal intestine, washed ready for reuse. Alternatively, they could buy secondhand condoms, which had already been washed and hung up to dry.

Modern condoms are usually made from latex, which wasn't invented until 1920, but for centuries people have used more rudimentary condoms, made of other materials. Condom use may even go back as far as the Roman Empire and beyond, although the evidence for this is sketchy. The first unambiguous reference to condoms was made in 1564 in Gabriele Falloppio's *De Morbo Gallico* (meaning "On the French Disease"—which was syphilis). Falloppio claimed to have invented a condom made of linen, to be tied onto the penis using a ribbon. Around this time, condoms were also being made from the intestines and bladders of animals such as sheep and goats. Archaeologists have even found surviving examples of condoms made from animal intestine dating back to the 1640s, at Dudley Castle in the West Midlands.

As the deadly syphilis epidemic spread throughout Europe, condom use became increasingly widespread. Casanova made a diary entry in 1753 describing his practice of buying condoms by the dozen. To test their quality, he would blow them up before use to make sure there were no holes. James Boswell and the Marquis de Sade also made references to condoms, and prophylactics can be seen in the background of a number of contemporary paintings, hanging up to dry, ready for their next sortie.

To make a sheep-gut condom, the condom-maker would soak a sheep's intestinal caeca in water for a number of hours, before turning it inside out, and leaving it to soften in a mild alkaline solution. Once the intestine was adequately softened, they would scrape away the mucus membrane, leaving the peritoneal and muscular coats, which were then treated with the vapor of burning sulfur. After this, the membrane would be washed, inflated, dried, and cut to a length of around seven or eight inches. Finally, the open end would have a ribbon attached, with which to tie the condom onto the base of the penis. Generally, a user would soak the condom in water before use, to make it more supple.

By the late eighteenth century, Londoners had a range of condom shops to choose from, and in this era before the prudish age of Victoria, these businesses would openly market their wares. One of Miss Jenny's more upmarket rivals was a Mrs. Phillips, who ran a comprehensive sex shop in what is now Covent Garden, selling condoms, sex books, flagellation machines, and "widow's comforters." Mrs. Phillips's advertising handbills boasted of her thirty-five years' experience in making and selling her "implements of safety," which even came with ribbons in regimental colors. Her pamphlet closed with a short verse:

> *To guard yourself from shame or fear*
> *Votaries to Venus, hasten here;*
> *None in my wares e're found a flaw,*
> *Self-preservation's nature's law.*

what is a ubi?

There can't be many of us who haven't wondered at some point quite what was written on our illegible prescription or hospital chart. Doctors are educated people, skilled with their hands; so why can't they write clearly? Well, one reason might be that they don't want us to know what they're writing, as it seems that sometimes, the joke is on us.

For example, if you ever do see the letters UBI on your medical records, you might like to know that your doctor clearly believes that you are suffering from an Unexplained Beer Injury. Doctors and nurses have a whole range of amusing acronyms and slang terms with which to express their true opinions about our often self-inflicted ailments, although more open access to patient notes nowadays means that these terms are more likely to be used in conversation and private notes than on a patient's formal records. Here is a selection:

Plumbum oscillans—Latin for "swinging the lead," implying that the patient is trying his luck for a medical excuse

DBI—Dirtbag Index, calculated by multiplying the number of the patient's tattoos by the number of missing teeth

CTD—Circling the Drain

GPO—Good for Parts Only

Departure Lounge—Geriatric ward

LOBNH—Lights On but Nobody Home

Oligoneuronal—A fancy term for "thick"

GOK—God Only Knows

BTSOOM—Beats The Shit out of Me

PAFO—Pissed and Fell Over

HTK—Higher than a Kite

TTFO—Told to Fuck Off

FOS—Full of Shit: a patient not considered to be entirely honest

FLK—Funny-Looking Kid

AGMI—Ain't Gonna Make It

Code Yellow—Bladder control emergency

Code Brown—Must I spell it out?

DFKDFC—Don't Fucking Know, Don't Fucking Care

what was *the fruits of philosophy?*

It was a scandalous book, the *Lady Chatterley's Lover* of its day. It was written by Charles Knowlton, a fairly unremarkable doctor in Ashfield, Massachusetts, who had literary ambitions. In 1832 he wrote a small pamphlet, *The Fruits of Philosophy, or the Private Companion of Young Married People*, which he showed to a few of his patients. Unlike *Lady Chatterley's Lover*, it was not a novel. Rather, it was simply a nonfiction guide to conception, with advice concerning infertility and impotence. It also contained tips for birth control, including a method for douching the vagina after intercourse.

Then as now, there were some people who felt that the idea of birth control was irreligious. A campaign was begun by the town's minister, Mason Grosvenor, and Knowlton was prosecuted and fined. The book was then reprinted in Boston, and Knowlton was punished again, this time given three months' imprisonment and hard labor. After his release, Knowlton continued with his medical practice, and developed an excellent reputation as a reliable and conscientious physician. He died of heart failure in 1850.

Twenty-seven years after Knowlton's death, the book caused a fresh sensation in London. It had been published by the provocative freethinkers

Charles Bradlaugh and Annie Besant, who were consequently prosecuted under the Obscene Publications Act of 1857. The solicitor-general was outraged by the book:

> *I say this is a dirty, filthy book, and the test of it is that no human being would allow that book to lie on his table; no decently educated English husband would allow even his wife to have it. The object of it is to enable persons to have sexual intercourse, and not to have that which in the order of Providence is the natural result of that sexual intercourse.*

Bradlaugh and Besant were both given six months in prison, along with a $300 fine, although these were quashed on appeal, thanks to a technicality. The trial, however, had the effect of turning the book into a sensation, which sold more than 150,000 copies. The book's resulting fame is believed to have had a major, positive impact on the use of contraception on both sides of the Atlantic, and it is even credited by some with reversing Britain's population growth.

what was the "french pox"?

The answer is syphilis, at least according to the English. Syphilis was known by different names all over Europe, in a manner that seemed to follow a clear pattern: Each nation blamed syphilis on its worst enemy. In France, it was *le mal de Naples*, while the Italians called it the Spanish disease. The Japanese called it the Chinese disease, while most of China blamed the Cantonese, those people from the province of Guangzhou.

In fact, syphilis is just one example of a broader trend. In England, the French were seen as being crude and sexually permissive, so condoms became known as French letters, genital herpes was known as the French disease, and an open-mouthed kiss was a French kiss. Because of this shady reputation, French postcards and novels meant pornography, and you would apologize for swearing by asking someone to "pardon my French."

In France the same thing happened, only the other way round. On that side of the channel, *filer à l'anglaise* means to leave without permission, or without saying good-bye, literally "English leave." In England, the exact same concept is called "French leave." Our French letter is their *capote anglaise*, while French kissing becomes *le baiser anglais. Plus ça change* . . .

what was the public's great fear, following the discovery of x-rays?

X-rays were discovered by accident, by a German physics professor named Wilhelm Röntgen in 1895, who was experimenting with electrons in vacuum tubes. Waving his hand between the covered tube and a screen, he saw an image of the bones of his hand projected onto the screen. He then took an X-ray picture of his wife's hand, and this astonishing image caused an international sensation.

X-rays are a form of radiation that passes through different parts of the body in different ways, depending on their density. As a result, X-rays can produce images of the internal structure of the human body, and these images can be captured on photographic plates. This discovery was a huge breakthrough, as it meant that for the first time doctors were able to look inside the human body without surgery. Within weeks of Röntgen's find, doctors were taking pictures of broken bones, gallstones, lesions, and all sorts of objects lodged inside people's bodies, including bullets and needles.

Today we know that, while useful, X-rays can

increase the risk of cancer, and therefore need to be used sparingly. In 1895, however, no one knew any of this. Instead the public's chief fear, it seems, was that X-rays could be used to look through people's clothes, to ogle their naked bodies. A London company began marketing X-ray-proof underwear, and is said to have made a small fortune. The New Jersey state legislature even brought in a law forbidding the use of X-rays in opera glasses, on the grounds of protecting public modesty!

Interestingly, while these fears may now seem rather quaint, today we face a very similar issue. Controversial new scanners have been introduced at airports, which do clearly show the naked bodies of passengers, including their genitals. Airport staff have already been caught misusing the scanners— a male member of staff at Heathrow Airport was alleged to have used the scanner to take a photograph of an unwitting female colleague, who then complained of sexual harassment. At Manchester Airport, two women were barred from boarding their flight after refusing to walk through the scanner. At Miami International Airport, a fight broke out after a male member of the staff was taunted for the size of his penis, as revealed by the scanner. Perhaps in the future these scanners will seem as harmless and trivial as X-rays, but at present it seems unlikely.

who was patient zero?

In 1987, a gay investigative journalist named Randy Shilts published *And the Band Played On*, a book about the early years of the AIDS epidemic. The book became a highly controversial bestseller, particularly for its most sensational reported claim: that one highly promiscuous gay man, the so-called Patient Zero, had played a major role in bringing AIDS from Africa to the United States, and then spreading it from city to city. Patient Zero was named in the book as Gaëtan Dugas, a Canadian flight attendant who had died of AIDS-related illnesses in 1984.

Most of the book's marketing and coverage focused on the Patient Zero angle. *Time* magazine opened its review with the headline, "The Appalling Saga of Patient Zero." The *National Review* called Dugas "The Columbus of AIDS." *California* magazine promoted its serialization of the book with an ad showing a picture of Dugas over text that ran: "The AIDS epidemic in America wasn't spread by a virus. It was spread by a single man. . . . A Canadian flight attendant named Gaëtan Dugas . . ."

According to the book, Dugas was blond, handsome, charming, and extremely promiscuous. He would travel the world, picking up men everywhere he went. Arriving at a gay bar, he would survey the room, announcing with satisfaction, "I'm the prettiest one." After developing Kaposi's sarcoma, which is a form of AIDS-related skin cancer, he was warned by doctors that he might infect others, but he carried on regardless, refusing to stop having unprotected sex. Sometimes after an encounter, he would gesture toward the purple lesions on his chest, telling his potentially doomed partner, "Gay cancer. I have it, maybe you'll get it too."

AIDS was a terrifying and emotive subject in 1987, and Shilts's book had the effect of personalizing the disease and allocating blame onto one man, and also implicitly onto a whole community. The account given of Dugas reinforced existing prejudices and stereotypes that already perceived gay people as promiscuous and irresponsible. AIDS now had a face, a Typhoid Mary for the 1980s, who confirmed the view that this was a disease spread and ultimately caused by the gay community.

This account of Dugas was not just unreasonable in the way it sought to apportion blame, it was also factually incorrect. Shilts's account was based on an earlier study, which had looked for links among a number of the first AIDS sufferers and found that forty of them were directly or indirectly linked to Dugas. No one was named in the study, and Dugas was referred to only as "Patient O," with the O standing for "Out of California." The name "Patient O" was misunderstood as "Patient Zero," implying that Dugas was the first person to bring the disease to the United States. Furthermore, the suggestion that it was Dugas who had infected these men with HIV was also unlikely to be true, as AIDS can take years to manifest symptoms, whereas most of these cases were diagnosed within months of the relevant encounters.

A more recent study in 2007, which looked at HIV gene sequences in the blood samples of early AIDS sufferers, found that Dugas could not have brought AIDS to America, as it had already arrived from Haiti, possibly as early as 1969. AIDS had already been circulating in the United States for around twelve years before it was first identified in 1981.

what was antonie van leeuwenhoek the first man to see?

Antonie van Leeuwenhoek was a Dutch tradesman and scientist, born in 1632, who today is generally regarded as the father of microbiology. His career would have been an extraordinary one for any scientist, so it is even more remarkable that van Leeuwenhoek was really an enthusiastic amateur. He had no university education, and no scientific training. His family was not wealthy, so he worked in a range of mundane jobs. And yet, through his skill and intellectual curiosity, he made some of the most astonishing discoveries in the history of science.

Perhaps inspired by the polymath Robert Hooke, van Leeuwenhoek began making simple microscopes sometime before 1668 and using them to make observations of anything and everything. Over the course of his life, he is known to have made over five hundred different microscopes, although fewer than ten of them have survived to the present day. Van Leeuwenhoek's microscopes were not the compound microscopes used today, which use multiple lenses; instead they were more like high-powered magnifying glasses. Yet his skill at molding glass, grinding lenses, and managing the levels of light meant that his microscopes could achieve magnification as high as 275 times, producing brighter and clearer images than any of his contemporaries' instruments.

Not only was van Leeuwenhoek a technical pioneer, he was also determined to explore the new worlds that these tools could reveal. He made lots of observations, becoming the first man in history to see microscopic life forms including bacteria, protists, and nematode worms. He discovered blood cells, the banded structure of muscle fibers, and he was also the first man to ever see his own spermatozoa swimming. He wrote numerous letters to the Royal Society in London, detailing his techniques, theories, and discoveries. He was initially met with some skepticism, as this microscopic world had simply never been seen before, but his findings were verified by his hero Hooke, among others. In

1680, van Leeuwenhoek was elected as a full member of the Royal Society.

Despite van Leeuwenhoek's incredible findings, microbiology was largely neglected after his death. There was no theory at the time that in any way connected these tiny microorganisms with disease, as the prevailing wisdom was still the Aristotelian idea that disease was caused by miasmas and bad air. As a result, the baton passed on by van Leeuwenhoek was not taken up again for more than a century after his death, until the advent of Louis Pasteur and germ theory.

chapter nine

public health

By medicine life may be prolonged, yet death will seize the doctor too.

WILLIAM SHAKESPEARE

why did the panama canal take more than thirty years to build?

The Panama Canal was one of the largest and most challenging engineering projects ever undertaken. The canal links the Atlantic and Pacific Oceans, and is one of the busiest shipping routes in the world. The project was first begun in 1880 by France, but yellow fever and malaria decimated the workforce, killing vast numbers of laborers. Panama is a humid, tropical environment, ideal for mosquitoes, which transmit both malaria and yellow fever.

In response, the French invested heavily in modern hospital facilities, but no measures were taken against mosquitoes, as no one yet knew that they were the vectors for the diseases. Most buildings had no mosquito screens or nets, and hospital beds often had pans of water around their legs, which provided a perfect environment for mosquito larvae to grow. In 1893, the project was abandoned after more than 52,000 workers had fallen sick, and perhaps as many as 20,000 had died.

In 1903, the project was restarted by the United States, which bought out the French equipment and rights. The U.S. approach was much better organized, under the leadership of engineer John

Frank Stevens. The Panama Railway was rebuilt to support the construction, and proper accommodation was built for the workers.

However, the most important change was a fresh approach to the problem of disease. In the intervening years, Sir Ronald Ross in India had conclusively demonstrated that malaria was transmitted by mosquitoes. Now a war against the pests was fought by army doctors Major Walter Reed and the handsome-sounding Colonel William Gorgas. They oversaw a hugely ambitious and expensive eradication program. Mosquitoes lay their eggs on the surface of standing water, so all pools were either drained, or sprayed with oil and insecticide. Small streams would have a dripping oil bucket placed over them, creating a film of oil over each still patch of water, killing the larvae. All canal workers were given free quinine, and all homes were fumigated, using up the United States' entire stocks of sulfur and pyrethrum. Anyone who became infected was quarantined, which meant they were kept in screened cages that mosquitoes could not get into, thus preventing them from reinfecting. The campaign was enormously expensive—it has been estimated to have cost as much $10 per mosquito killed.

Still, it worked. The canal was opened in 1914, two years ahead of schedule. In 1906, there was only one case of yellow fever among the construction workers. From 1907 to 1914, there were none.

Malaria was more persistent, but between 1906 and 1909, the death rate from malaria fell dramatically among employees, from 11.59 per 1,000, to 1.23 per 1,000. Over the period of U.S. control (1904–1914), there were only 5,609 fatalities, which marked a huge improvement. Today, the canal zone is free from both yellow fever and malaria.

who was typhoid mary?

Mary Mallon was the first person in the United States to be diagnosed as a healthy carrier of typhoid. She was born in County Tyrone, Ireland, in 1869. After immigrating to New York in her teens, she became a cook, working for a series of well-to-do families. Everywhere she went, people seemed to fall sick with typhoid, but she had never had the disease herself, so there seemed to be no reason to think that she could be responsible.

In 1906, she travelled to Long Island with the family of banker George Warren, to spend the summer cooking for them. After a few days of eating Mary's food, the family started to fall ill. An investigator, George Soper, was called, and he soon determined that the likely cause was Mary herself. Soper knew about the possibility of a person being a healthy carrier, meaning that they might carry a disease without ever having suffered from it themselves, and then transmit it to others who may suffer the full effects.

In the case of typhoid, a healthy carrier may have suffered only mild flu-like symptoms when they first became infected, and then survived to pass on this fatal disease to others.

Typhoid fever has been a major killer for centuries, and it is still endemic in much of the world. It is caused by the bacillus *Salmonella typhi*, which can be transmitted in human urine and feces. Symptoms include sudden and prolonged fever, nausea, diarrhea, and headaches. Typhoid is usually passed on through contaminated food or water, particularly if the person preparing the food doesn't wash his or her hands after going to the bathroom. When questioned, Mary stated that she rarely washed her hands, as she felt there was no need.

Soper tried to approach Mary, to request samples of her urine and feces for testing, but she reacted furiously, and lunged at him with a carving fork. After a few more attempts were met with similar aggression, the authorities were forced to intervene, and Mary was forcibly quarantined and kept in a cottage on Brother Island on the East River. This presented something of a legal and ethical conundrum, as Mary was being held prisoner without trial, against her will, having committed no crime. The concept of a healthy carrier was a relatively recent discovery, and there was no law that covered it. The Greater New York Charter did have a provision to prevent a person from willfully infecting others, if that person posed a threat through being "sick with any contagious, pestilential or infectious disease," but of course although Mary certainly did pose a threat, she wasn't actually sick.

While incarcerated, Mary continued to protest and appeal, as public interest grew in the case grew. Eventually she was released, on the condition that she sign an affidavit vowing never again to work as a cook. She complied with this, but soon disregarded it once she was freed. Mary was single and child-less, and none of the other jobs available to her paid as well as cooking did.

In 1915, typhoid broke out at the Sloane Mater-nity Hospital in Manhattan. Twenty-five people were infected, and it soon became clear that the source was a cook named Mrs. Brown, which was a pseud-onym Mary had begun using. When Mrs. Brown's true identity was eventually revealed, public sympa-thy for Mary quickly drained away. She now knew about healthy carriers, even if she didn't believe in them, and her use of a false name added to the impression that she was willfully risking other peo-ple's lives. Eventually, Mary was quarantined again, and she remained incarcerated for the rest of her life, before eventually dying of pneumonia in 1938. The autopsy found that she was still infected with typhoid bacteria on the day of her death.

Over the course of her career, Typhoid Mary is known to have infected at least fifty-three people, killing three, but the numbers may well be much higher. Her name lives on into the computer age, as a "Typhoid Mary" is now a slang term for a computer user who recklessly refuses to install adequate virus and malware protection, and consequently endangers other computer users in their network.

what does vitamin e do?

Bizarrely, no one really knows, and it may even do nothing, or at least nothing useful. Vitamin E is found in a wide range of foods, including nuts, seeds, wheat, eggs, spinach, avocado, and asparagus. In the past, the vitamin was thought to be an antioxidant that protected cells from aging by combating free radicals. There were hopes that high doses of vitamin E could help to prevent heart disease, cancer, strokes, and Alzheimer's disease.

Recent clinical studies have been unable to demonstrate any of these benefits, however, and in fact have suggested that high doses of vitamin E may be actively harmful. In 2005, the Heart Outcomes Prevention Evaluation trials, which studied more than 10,000 patients fifty-five years old and over, found no heart benefit from taking high doses of vitamin E, and in fact the vitamin seemed to be linked to a higher risk of heart failure. Other studies have also failed to find any benefit for vitamin E in preventing heart disease.

The Heart Outcomes trial also looked at cancer, and found no difference between those taking

vitamin E and the control group. Other recent studies have looked at cancer in women, and prostate
cancer in men, and also found no benefit to taking
vitamin E. A National Cancer Institute study into
lung cancer found that those taking vitamin E were
at a slightly higher risk of developing the disease.
Other studies have looked into possible links with
Alzheimer's disease, but again found no benefit.

If anything, vitamin E may be actively harmful,
if taken in high doses. A meta-analysis conducted
by Johns Hopkins University in 2004 found that
the risk of dying within five years rose by about
5 percent in people taking at least 400 international
units (IU) of vitamin E per day. One possible explanation for this is that vitamin E is an anti-coagulant
that prevents blood clotting.

why were low-lying
towns particularly
susceptible to cholera?

After John Snow's conclusive demonstration that
the Soho outbreak of cholera was caused by the
contaminated Broad Street pump, one might imagine that the case for cholera being transmitted by
water would have been resolved. Instead, there was
considerable doubt about Snow's findings. Many
scientists still believed that diseases were primarily
caused by miasmas, and cholera was no exception.

They felt that Snow's epidemiological map proved nothing, because the concentration of cholera cases around the pump only demonstrated that the ground by the pump, or somewhere near it, was the source of bad air, not polluted water.

In fact, Snow had already made a convincing case against such objections. His study had included a number of cholera victims who had taken water from the Broad Street pump, but who did not live in the area, and therefore would not have been affected by bad air. The most compelling example was a woman who lived in Hampstead, an area that was untouched by cholera, who never went to Broad Street herself, but who was regularly supplied with a bottle of water from the area, because she had lived there in the past, and liked the taste of the water. She developed cholera, as did her niece who drank from the same bottle.

Snow's chief opponent was William Farr, superintendent of the statistical department at the General Register Office, who was widely recognized as the leading authority on the use of statistics in the study of disease. Farr had produced a persuasive study comparing the rate of cholera according to elevation from sea level. Amazingly, Farr's study seemed to clearly show that the nearer a person lived to sea level, the more likely he or she was to contract cholera. This, Farr argued, was clear evidence that the source of infection was something in the ground.

Snow responded by pointing out that the most elevated towns in Britain—Wolverhampton, Dowlais, Merthyr Tydvil, and Newcastle-upon-Tyne—had all suffered enormously from cholera. But many low-lying sites that happened to be supplied by a well, such as the Queen's Prison, Bethlehem Hospital, and Horsemonger Lane Gaol, had all largely remained untouched, despite being surrounded by areas that were badly affected.

Snow's explanation demonstrated that in fact the issue was not elevation from sea level, but where towns drew their water. There was some correlation between elevation and cholera, because low-lying towns tended to draw their water from tidal rivers, which meant that their water supply was liable to become infected by sewage traveling upstream.

By 1866, Farr had become a convert to the cause of germ theory. He produced a paper demonstrating the increased mortality rate for people who drew their water from the Old Ford Reservoir in East London. Thanks to the work of Edward Jenner and Louis Pasteur, germ theory was becoming widely accepted, and major cities were building new facilities to collect and treat sewage, a process that ultimately resulted in cholera being eradicated in Western cities.

was the "father of jogging" killed by jogging?

Jim Fixx was the key figure in inspiring the jogging boom that began in the United States in the late 1970s, and arguably continues to this day. He wrote several best-selling books, including *The Complete Book of Running*, which had a major impact on its release in 1977, selling more than a million copies. Fixx, who has been described as the "founding father of jogging," had started running in 1967, when he weighed 240 pounds and was a heavy smoker. He credited jogging with turning around his health, helping him to lose 60 pounds and give up smoking, and extending his life.

On July 20, 1984, Fixx died of a heart attack, at the age of just fifty-two. He was jogging at the time, which has led to the widespread belief that his death was caused by jogging, and that therefore the health benefits claimed for jogging are illusory. Famously, comedian Bill Hicks, who had little time for health fanatics, developed a routine mocking Fixx and relishing the irony of his death.

Yet the evidence suggests that Fixx's death was not in fact caused by jogging. His family had a history of heart trouble; his father had suffered a heart attack at age thirty-five, and had died of a second one at forty-three. Jim Fixx's autopsy showed that he died of blocked arteries: Three of his coronary arteries were blocked with cholesterol, suggesting that a more likely risk factor for his death was diet, not jogging, perhaps linked to his earlier, unhealthy years before he took up jogging.

It is true that the risk of heart attack goes up slightly during exercise, but this risk is much lower for people who exercise regularly, and pales into insignificance compared with the enormous benefits that exercise brings. In just one example, numerous studies have shown that regular exercise reduces the risk of a heart attack by between 50 and 80 percent. The likelihood therefore is that, while he may have died relatively young, Jim Fixx's life was nonetheless extended by jogging.

why are pig farmers more likely to have their appendixes removed?

In 1991, a Finnish scientist called Markku Seiru conducted a fascinating study into pig farmers and abattoir workers in Finland, comparing the rate of appendectomies among this group with that of the general population. The study found that pig farmers were around 2.5 times more likely to have had their appendix removed, compared with the norm, while abattoir workers were almost 4 times more likely to have had an appendectomy. But why should working with pigs carry such a bizarre and specific risk?

The answer is that a high proportion of pigs carry a bacterium called *Yersinia enterocolitica*, which can contaminate food, and causes a disease called yersiniosis. Yersiniosis is usually caused by eating undercooked meat, unpasteurized milk, or contaminated water. The symptoms include diarrhea, fever, and abdominal pain, and it is often confused with appendicitis. If an infection is properly diagnosed, doctors will often allow it to run its course, while more severe cases may be treated with antibiotics.

Yet because these symptoms often resemble appendicitis, it is not uncommon for patients to be given unnecessary appendectomies. At the time of the Finnish study, around 35 percent of pigs in Finland were found to carry *Yersinia enterocolitica*, which presumably explains the high rate of appendectomies among farm workers. Because this group were much more likely to encounter yersinia bacterium, they were therefore also more likely to have their symptoms confused with appendicitis, leading to unnecessary appendectomies.

why did organ donor numbers drop dramatically in 1978?

In 1978, there was a sudden, dramatic drop in the number of potential organ donors coming forward, with the reduction reported as being as high as 60 percent. Doctors were puzzled at first, as there seemed to be no obvious reason why rates should drop so suddenly. Then, a compelling theory emerged, which suggested that the cause of this decline was the best-selling novel *Coma*, written by Robin Cook, which in 1978 was adapted into a hit film directed by Michael Crichton. *Coma* was a creepy medical thriller that told the story of an unscrupulous hospital where patients were deliberately killed so that their organs could be harvested and sold to rich clients. Both the book and the film made a big impact, and after their release, rates of organ donation dropped dramatically.

Fortunately, the public's sensitivity to stories about organ donation can also be a force for good. In 1982, the parents of eleven-month-old Jamie Fiske were desperate. Their daughter had been born with biliary atresia, a congenital liver condition that at the time usually led to death before the age of four. The only hope was a liver transplant, but there were no suitable livers available. Luckily, Jamie's father was a well-connected hospital administrator, and so he began a publicity campaign, contacting thousands of doctors, and recruiting major public figures such as Senator Edward Kennedy and news anchor Dan Rather to lobby on Jamie's behalf. The campaign worked. Jamie Fiske became a major news story, and the families of five hundred potential donors contacted the family, offering to help. A suitable liver was found, and Jamie survived. Thanks to this publicity, requests for organ donor cards went up by more than 30 percent, and the number of families allowing doctors to remove organs from relatives who suffered irreversible brain damage went up from 50 percent to 86 percent.

are we living longer?

The simple answer is yes, although it is slightly more complicated than it might first sound. Average life expectancy has obviously been going up for some time, with a dramatic rise since the nineteenth century. In medieval Britain, average life expectancy was around thirty years. This figure is unlikely to have climbed beyond forty at any point until the start of the twentieth century. Over the course of the twentieth century, life expectancy in many parts of the world doubled. In the United States the current life expectancy at birth is seventy-eight, and in many parts of the world it is over eighty. These increases are believed to be largely due to improved sanitation and clean water, as well

as improvements in medicine, nutrition, childbirth, surgery, and the development of vaccines.

The definition of life expectancy may warrant
some explanation. In ancient Rome, for example,
the average life expectancy at birth was twenty-five,
but this is not to suggest that most people died at
around twenty-five. Twenty-five was an average, taking into account the huge numbers of people who
died as babies, in a period when infant mortality
was extremely high; or who died from childhood
disease, or military conflict, and so on. Anyone who
managed to survive their early years would see their
life expectancy increase dramatically, even as they
got older. Thus, while the average life expectancy
in Roman times may have been twenty-five, many
people did live on to what we would consider a ripe
old age.

In fact, throughout history, people have lived to
ninety and beyond, and perhaps even beyond hundreds. In ancient Egypt, the Sixth Dynasty pharaoh
Pepi II is believed to have lived beyond one hundred. Pharaoh Rameses II was recorded as having
lived to about ninety, and tests on his mummified
tomb seem to confirm this. In ancient Greece, the
philosophers Pyrrho, Eratosthenes, and Xeno-
phanes all reached their nineties. Socrates died
just before his ninetieth birthday, and only because
he was executed. The astronomer Hipparchus of
Nicea is said to have lived to 109. In the fourth century a number of Christian priests lived into their

nineties. Michelangelo was still working at eighty-nine, and a number of Native American chiefs lived into their hundreds.

In other words, at many points throughout human history there were lots of people living who would outlast the vast majority of people living today. The oldest person to have lived was Jeanne Calment of France (1875–1997), who died at age 122. This suggests that, despite all our advances in medicine, sanitation, nutrition, and health, the maximum human life span doesn't seem to have increased by a huge amount, even though the average life span certainly has. This is demonstrated even today, in parts of the world where medicine has had only a very limited impact. The lowest life expectancy rates in the world are found in Africa, and yet Africa also produces centenarians. In regions of the Caucasus, there are said to be much higher rates of centenarians than in the United States, and medicine has presumably had only a limited impact there too.

Another complication is that, while average life expectancy has clearly increased dramatically, it's not clear how much longer this trend will continue, and some suspect that it may already have peaked. Demographers agree that there will be many more centenarians in the decades to come, and yet it is also possible that life expectancy at birth may be beginning to drop, at least in the United States. Today, a much higher proportion of the population is overweight or obese, particularly among the young, and many people tend to lead sedentary lives, with little exercise. The majority of deaths in advanced Western countries are caused by heart disease, cancer, diabetes, and stroke, and those with sedentary lifestyles and high BMIs are more at risk from these diseases. As a result, a recent report suggested that average life expectancy could drop by as much as five years, based on current rates of obesity.

who was patient x?

Patient X was the name given to a former U.S. Marine who preferred to keep his true identity a secret after inadvertently taking part in what was to become an award-winning medical study. He was bitten on the lip by his pet rattlesnake, and insisted on curing the venomous injury himself using electroshock therapy. He did this by attaching the spark plug wires from his car to his lip, and revving the engine to 3,000 rpm for five minutes.

Thanks to his brave experiment, Patient X was the joint winner of the 1994 Ig Nobel prize for medicine. He shared the prize with Dr. Richard C. Dart of the Rocky Mountain Poison Center and Dr. Richard A. Gustafson of the University of Arizona Health Sciences Center, for their "well-grounded" medical report "Failure of Electric Shock Treatment for Rattlesnake Envenomation." After receiving his award, Dart commented, "I was stunned to receive the 1994 Ig Nobel Prize in Medicine, although not as shocked as our patient."

The Ig Nobel awards are a series of light-hearted science awards, whose stated aim is "to first make people laugh, and then make them think." In 2009, the winner of the medicine prize was Donald L. Unger of California, who had cracked the knuckles of his left hand twice a day for fifty years, while rarely cracking the knuckles of his right hand, to test whether or not knuckle-cracking causes arthritis. He found that after fifty years of this experiment, there was no arthritis in either hand, and no difference between the two hands, and thus concluded that knuckle cracking does not cause arthritis.

did the new yankee stadium have to fit wider seats, to accommodate the increasing girth of fans?

Indeed it did. The original Yankee Stadium was built in the 1920s, when the average American was considerably thinner. At that time, seats were just 15 inches (38 cm) wide, but as the American population has grown steadily wider, this proved to be uncomfortable for all concerned. Since 1980, obesity rates in the United States have doubled, and today around 65 percent of Americans are overweight or obese. In response to this trend, the owners of Yankee Stadium removed nine thousand seats, and replaced them with seats that were considerably larger, at 19 inches (48 cm) wide. In 2009, a brand-new Yankee Stadium was opened, just across the street from the old site. Again, seat sizes were increased. In the new stadium, seat widths range from 19 to 24 inches (48 to 61 cm).

Nor is this a problem restricted to the United States. In the UK, more than half the population are now overweight or obese, and the rate of obesity has increased almost fourfold in the last twenty-five years. In London, plans for the 2012 Olympic stadium had to be redrafted to accommodate the ever-increasing size of British bottoms. In the original plans, all 20,000 seats were to be 18 inches (46 cm) wide, but they will now be increased by 2 inches (5 cm). British amusement parks have also started offering special rows of "outsized" seats for larger visitors.

This issue has been a particular problem for airlines. Airlines will generally try to cram in as many seats as possible, to maximize their earnings, and keep prices low. There was a recent outcry when

British Airways tried to cram another line of seats into the economy section of its Caribbean flights, making each row ten seats wide, rather than the usual nine. In the end, the company received so many complaints that it reversed the plan. In 1996, the Civil Aviation Authority published a report recommending a minimum seat width of 19.6 inches (50 cm), but some airline seats are just 16.2 inches (41 cm) wide.

As a result, many obese passengers simply can't fit into a single seat. Many airlines now either encourage or require larger passengers to buy a second ticket, so that they can sit on two seats, which has caused some controversy. In February 2010, filmmaker Kevin Smith reacted furiously when he was removed from a Southwest Airlines airplane prior to departure, apparently because he was considered too large to fly safely in just one seat.

Overweight passengers feel that this kind of policy is discriminatory. Others argue that it's only fair that each passenger should pay for the space they take up, citing horror stories of thin passengers being crushed by portly co-passengers. In 2002, Virgin Atlantic paid £13,000 (equivalent to around $20,000) in compensation to Barbara Hewson, of South Wales, who suffered sciatica, torn leg muscles, and a blood clot after being crushed by an obese neighboring passenger on a transatlantic flight.

why are there so many contradictory health stories in the news?

Every week, the news seems to bring us a new health breakthrough, and as often as not it completely contradicts a different health breakthrough we read about only a few months ago. These stories often focus on food. Chocolate is an excellent source of antioxidants, the headline reads, until next week it is a key factor in childhood obesity. Red wine lowers the risk of heart attack, the next week it causes liver disease. Coffee lowers the risk of heart disease, but caffeine can increase blood pressure, increasing the risk of a heart attack. Oily fish can boost brain function and IQ, unless they contain poisonous mercury, increasing the risk of heart disease. Soy milk is healthier than cows' milk, unless it increases the risk of breast cancer. And so on.

However, as nice as it would be to blame all our ills on one type of food, or alternatively to rely entirely on one magic superfood, there seems to be no one single factor which can guarantee good health. The best way to protect your health is simply to follow the basic rules that we all generally know already: avoid smoking, take regular exercise, eat a balanced diet including fruit and vegetables, drink in moderation if at all, and get plenty of sleep.

So why do these contradictory health stories keep appearing in the press? The main reason is that they make good stories, which people love to read. They tend to contradict one another simply because the studies these stories are based on rarely draw the kind of dramatic conclusions that the media claim for them. Studies of this type are often run on a small scale, and are merely intended to test a possible mechanism before more detailed research takes place. A hypothetical case-control study might look at coffee drinkers, for example, and find that they have fewer heart attacks than the rest of the population. This doesn't prove that coffee reduces the risk of heart attack, it merely suggests that there might be a possible link, and that further studies can now take place. Hopefully, there will then be more studies, using control groups, and randomization to eliminate sources of bias, in an effort to make all things equal, except for the subjects' coffee drinking.

Even then, the study is bound to be unsatisfactory in some way. For a study to be really effective, it needs to sample a large number of people, with controls and randomization, and follow them for a long time—after all, the effect of coffee on the heart may take years to appear. Studies of this type are extremely expensive and difficult, and they carry ethical implications. If the researchers really believe that there is a likely link between coffee and heart attacks, is it justifiable to ask the control group to drink coffee for years, possibly endangering their health in the process?

Large-scale studies of this kind are rare, so health stories will often imply a causal link where none has been shown. For example, a number of recent studies have shown a possible small connection between eating processed pork and higher rates of colorectal cancer. The press has reported this hysterically with headlines such as, "Why eating just one sausage a day raises your cancer risk by 20 percent" in the *Daily Mail*. But correlation does not equal causation. There are many reasons why the kinds of people who eat processed pork might also suffer from higher rates of bowel cancer.

Sometimes, these stories are based on studies that never even appear. Newspapers will sometimes receive an exciting-looking press release, run a sensational story, but then the actual study never materializes in any scientific journal. For example, in February 2004 the *Daily Mail* ran a story about a forthcoming study that demonstrated that cod liver oil was "nature's superdrug," based on a press release from Cardiff University. A year later, however, no study had been published.

who is "frozen dead guy"?

In 1989, a Norwegian named Trygve Bauge brought the frozen corpse of his grandfather Bredo Morstøl to the United States, where he intended to keep it cryogenically frozen. Trygve was a passionate believer in cryogenics, also known as cryonics, which is the practice of freezing dead bodies to preserve them in the hope that science will one day be able to bring them back to life. Trygve was not only trying to save his grandfather for the future, but also planning to establish his own cryonics clinic in Nederland, Colorado, where he set up home.

Trygve's plans were in truth somewhat amateurish. If you imagine a cryonics institute, you're presumably picturing some kind of shiny, chrome and glass high-tech laboratory facility, with gleaming, carefully engineered caskets submerged in liquid nitrogen. Trygve's facility consisted of a run-down shed, with a coffin in a plywood box, filled with dry ice.

Sadly, Trygve had to give up on his plans when he was deported in 1993 for overstaying his visa. He left behind his mother, Aud, and of course his grandfather's frozen body. When the local authorities found out about Bredo's frozen corpse, which was now at risk of thawing, their first reaction was to pass a law banning the keeping of frozen bodies. This law, incidentally, also made it technically illegal for anyone to keep a frozen chicken in their freezer, but no one seemed to be all that bothered. Nonetheless, the passing of the law raised more local interest, and Bredo was starting to generate significant press coverage. The local authorities decided that an exception could be made for Bredo, even though the law had been specifically passed to deal with his body, and so he was allowed to stay. Trygve arranged for a local company to continue regularly replacing the dry ice, and another local company even built the family a new shed.

Over time, the town became famous for its unusual resident, and so decided to capitalize on this by setting up an annual celebration. The first full weekend of March became the "Frozen Dead Guy" festival, bringing five thousand tourists into this remote town for a series of bizarre festivities. There are coffin races, on a course that goes over the roof of Grandpa Bredo's shed. There are snowshoe races, and snow sculpture contests, and a Grandpa Bredo look-alike contest. There is Frozen Dead Guy ice cream, and champagne tours of the famous shed. There is also a dance, with the slightly juvenile name of "Grandpa's Blue Ball," which revelers attend dressed as corpses. Hardy visitors can attempt the "polar plunge" into an ice-cold reservoir, but you usually have to break through the ice first.

One of the downsides of setting up your cryonics

facility in a shed is that the technical challenges are significant. Grandpa Bredo's corpse does not seem to have been kept to the very highest cryogenic standards, and has almost certainly defrosted more than once. As a result, even if it were to become possible to revive a frozen corpse, there seems to be little chance of a successful revival for Colorado's very own Grandpopsicle.

what was a "london particular"?

This was a kind of thick fog, or rather smog, which affected London intermittently for around one hundred years, from the Industrial Revolution until at least the 1950s. *Smog* is a portmanteau word combining the words *smoke* and *fog*; smog was caused primarily by exhaust fumes and industrial byproducts, in particular the burning of coal.

A London Particular was a common and routine occurrence, which was also known as a pea-souper, in reference to the yellow tinge that the fog often took on, thanks to its high sulfur content. Amusingly, in a strange backward twist, pea soup became known as "London Particular," in reference to the famous smog.

Although pea-soupers were common, they were nonetheless enormously disruptive. An incredulous *Time* magazine article from 1951 describes an absurd scene on a London airport runway: The plane lands in a thick fog, and is ordered to stay in position, as visibility is too poor for the pilot to taxi back to the arrival terminal. A bus is sent out to pick up the passengers, but it gets lost in the smog. A truck is then sent out to find the bus, but it too gets lost. Soon, there are five separate search parties out searching, hopelessly groping around in the fog. Eventually, a man on a motorbike finds the plane, but realizes he now has no idea where the terminal is, so he rides off. Eventually, the passengers are rescued, but of course when they get to the terminal, their luggage has been lost.

The smog that appeared in December 1952 was unusually thick, but did not seem to be anything particularly out of the ordinary. But a number of unusual factors were about to combine to spectacular effect. The recent weather had been very cold, so people had been burning more coal than usual to keep warm. There was more pollution from exhaust fumes, as all the trams had recently been replaced with diesel-spewing buses. Prevailing winds from Europe had brought industrial pollution across the channel, which now became trapped in an anticyclone, with a layer of cold air trapped under a blanket of warm air.

As a result, this smog was thicker and longer lasting than it had been on any previous occasion. Visibility was limited to just a few yards, making driving practically impossible. Buses and ambulances stopped running, and the only public transport was the London Underground. As a result, there was no way for many people to get to the hospital. Amazingly, the smog even managed to find its way indoors, and concerts and film screenings were cancelled, as the audience couldn't see what was happening on the stage or screen!

After the smog subsided, things seemed to quickly return to normal. Yet soon there were strange signs. Undertakers ran out of coffins, and florists ran out of flowers and wreaths. There were no bodies littering the streets, but without anyone realizing it, the death rate had soared, to three or four times the normal level. Most of the fatalities were caused by bronchitis and lung infections. "Smog" was not usually listed as a cause of death, so statistics are hard to come by, but estimates suggest that as many as 12,000 people died as a result of just five days of smog, with around 100,000 more falling ill.

If any good can be said to have come from such a disaster, it is that the public outcry that followed the Great Smog led to new laws to protect the environment and limit air pollution. The City of London Act of 1954 and the Clean Air Acts of 1956 and 1968 restricted the kinds of fuel that could be used in industry, leading to a huge improvement in air quality. From our perspective today, as we face the potential calamity of climate change, examples such as this also set a powerful precedent. With the right combination of regulations and incentives, London Particulars, Los Angeles smog, and acid rain have all become outdated terms, raising hopes that one day we may feel the same combination of curiosity and nostalgia for phrases like "carbon footprint" and "greenhouse gas."

what is the cape doctor?

This is the name given by locals to the strong prevailing wind found at the Cape of Good Hope on the South African coast between spring and late summer. It is known as the Cape Doctor because for centuries the wind has been believed to carry pollution and pestilence away from Cape Town and out to sea. This wind can be extremely strong—it can blow double-decker buses over, and have people clinging horizontally onto lampposts for dear life.

It might strike you as odd to give a name to a wind system, but in fact there are lots of other prevailing winds from around the world that have specific names. You may for example have heard of the mistral, which is a strong, cold wind that blows from northern France down to the Mediterranean. There is also another wind doctor, the Fremantle Doctor, the name for the cooling sea breeze in Western Australia, which may have gained its name by blowing away the stench of burning corpses that was a feature of the ailing colony in its early years.

My personal favorite is the Williwaw, which is a sudden blast of wind descending from the coastal mountains at the Strait of Magellan, the waterway that divides the southern tip of South America from the archipelago of Tierra del Fuego. Historically, this was the main shipping route between the Pacific and Atlantic oceans, before the Panama Canal was built. Thanks to the unpredictable and powerful Williwaw, it was also one of the world's most dangerous stretches of water.

chapter ten

snake oil

God heals, and the physician has the thanks.

TRADITIONAL PROVERB

why is a quack called a quack?

A quack is someone who fraudulently claims to have medical expertise of some kind, and the term is often used to refer to the sellers of patent medicines. The name *quack* has nothing to do with ducks; it comes from the archaic Dutch word *quacksalver*, which means "a boaster who applies a salve." A quacksalver would thus quack about his salves, which is a pretty good definition of a traveling medicine man, selling his dubious nostrums at markets and county fairs.

does snake oil work?

When snake oil is mentioned today, it is generally used as a euphemism, a byword for fraudulent quackery. This use of the term refers to a bygone age, when patent medicine men traveled from town to town, selling snake oil, vegetable tonic, and Indian root pills. Their remedies rarely worked, and rarely even contained the ingredients they claimed to. The skill of these traveling "doctors" was all in the marketing, filled with extraordinary, bogus pseudoscientific claims and anecdotes. They would also frequently use shills, accomplices in the crowd who would pretend to be cured by the remedy in question. Today, we use the term "snake oil" to refer to pseudoscience, absurd health claims, and any kind of exaggerated salesmanship.

Even though the name "snake oil" has become synonymous with fraud, it seems that the product itself might, astonishingly, work. Snake oil has been used for centuries in traditional Chinese medicine, in which it is known as *shéyóu*. It has traditionally been used as a pain reliever and anti-inflammatory for joint pain, rheumatoid arthritis, and bursitis, and in fact it is still sold and used today. The snake oil in question comes from the Chinese water snake, a species found in China, Taiwan, and Vietnam.

Snake oil may have been brought to America by the gangs of Chinese laborers who worked on the Transcontinental Railroad, the railway that was built in the 1860s to link the east and west coasts. This was backbreaking work, and the Chinese workers are said to have rubbed snake oil on their aching joints to ease the pain.

We know today that Chinese water snakes are a rich source of eicosapentaenoic acid, or EPA, which is an omega-3 fatty acid. Omega-3 fatty acids are found in fish oil, and are believed to bring a range of health benefits. EPA is known to be an effective anti-inflammatory and pain reliever. In other words, despite its dubious reputation, snake oil might actually work.

This raises a rather tricky objection: If snake oil was one of the few patent medicines, or perhaps even the only one, that actually worked, how did it acquire such a bad name? The answer may be found in the specific composition. As mentioned above, Chinese water snakes have very high concentrations of EPA in their bodies; around 20 percent of their oil is EPA. The oil of American rattlesnakes, on the other hand, contains only 8.5 percent EPA. Even if the traveling hucksters were selling a product containing genuine snake oil, it would rarely if ever have come from a genuine Chinese water snake, so the odds are it would contain little or no EPA. One of the most popular American snake oil products was Stanley's Snake Oil, produced by Clark Stanley, the self-proclaimed "Rattlesnake King." Tests in 1917 found that Stanley's Snake Oil was made up of turpentine, camphor, and mineral oil—but no snake oil. In other words, genuine snake oil may have actually worked, but the product that was sold would very rarely have been genuine.

what was the "doctors' riot"?

Right now, you may be imagining an angry mob of men in white coats, wielding clipboards and stethoscopes, but in fact this was not a riot *by* doctors, but rather a riot *against* doctors. According to one report, the trouble started at the New York Hospital in April 1788, when a medical student named John

Hicks, Jr., waved the arm of a corpse at a group of nosy children who were peering into the dissection room. "This is your mother's hand," he shouted. "I just dug it up. Watch it or I'll smack you with it!"

The children ran off, but one boy was particularly upset, as his mother had only recently died. Could it really have been her arm? He went home and told his father what had happened, and they then went to inspect the lady's grave. Sure enough, the grave was empty, having recently been dug up, and whoever had done it had smashed open the casket, and hadn't even bothered to refill the grave with earth. The husband was furious, and vowed that someone would pay for this outrage. He gathered together a crowd of friends, and they marched through the streets of Lower Manhattan, to the New York Hospital, with the crowd swelling to hundreds on the way.

There was a reason for this outpouring of public outrage. Grave robbing was a growing problem, as medical students and anatomy teachers needed fresh corpses for their lessons. They usually only took corpses from the graves of the poor, homeless, or black, but in times of greater demand, any grave would do. The problem had become so bad that wealthy families would post armed guards at the graves of their relatives for two weeks after burial. After two weeks, the body would be too decayed to be any use for dissection.

The angry mob surrounded the hospital, blocking all the exits, and baying for the doctors' blood.

Luckily, most of the doctors had already managed to escape through the hospital's rear windows. One stayed behind with three medical students to guard the precious instruments and anatomical specimens, but it was to no avail, as the rioters broke in and destroyed everything in sight.

The mob was still not satisfied, and so they headed out into the streets, looking for doctors to attack. One man was beaten to the ground, simply for being dressed in black. The home of Sir John Temple was completely destroyed, despite his having no connection to medicine—it seems that the rioters had misread "Sir John" as "surgeon."

The riot continued the next day, with the mob attacking Columbia College, destroying valuable medical specimens and tools. That evening, the epicenter moved to the Manhattan Jail, where a number of doctors were being held for their own safety. The militia guarding the jail was led by Baron Friedrich von Steuben, who refused to use force, until one of the rioters hit him with a hurled brick, at which point he changed his tune, crying "Fire, Governor, fire!" At least five of the rioters were killed, with many more seriously wounded. The doctors treated the injured, and the riot ended.

Weeks later, a law was passed in New York allowing for the dissection of hanged criminals, but this made no difference, as the demand was so high that grave robbing continued unabated until at least the 1850s. And although a riot against doctors might sound like a strange and isolated incident, in fact there were as many as twenty-five similar incidents of public outrage at medical schools throughout the United States up until 1884.

how did dr. alpheus myers's "tapeworm trap" work?

In 1854, Alpheus Myers, a rural doctor from Logansport, Indiana, filed a patent for an ingenious product, a "new and useful Trap for Removing Tapeworms from the Stomach and Intestines." The trap consisted of a spring-loaded cylinder of gold, platinum, or some other rustproof metal, about three-quarters of an inch (1.9 cm) long, and a quarter of an inch (.6 cm) in diameter, attached to a cord. The cylinder would contain the bait; according to the doctor, cheese would be an attractive bait for a hungry tapeworm.

Before using the trap, the patient would have to fast for a week. The idea of this was to starve the tapeworm, to force it to climb up into the stomach, or even the throat, in search of food. After fasting, the patient was to swallow the trap, leaving one end of the cord hanging from his mouth, and then wait for six to twelve hours. During this time, Dr. Myers predicted that the tapeworm would take the bait, causing the spring-mounted cylinder to close around it. Myers warned that the spring should be strong enough to grasp the worm, but not strong enough to cut its head off. Myers believed the patient would notice when the tapeworm became trapped, either because they would feel it, or because the worm would tug on the cord. Then, "The patient should rest for a few hours after the capture, and then by gentle pulling at the cord the trap and worm will with ease and perfect safety be withdrawn."

As odd as it may sound, the device may have been successfully used at least once. A year after the patent was granted, *Scientific American* reported that Dr. Myers had used the trap to remove a fifty-foot (15.25 m) tapeworm from a patient, "who, since then, has had a new lease of life."

what was "graham's celestial bed"?

James Graham was a self-styled sex guru, and one of the leading quacks of the eighteenth century. He was born in 1745 in Edinburgh, where he trained in medicine, although he never finished his degree. While traveling around America, where he presented himself as an eye specialist, he learned about electricity (from Ben Franklin, he claimed, but this was almost certainly false).

In 1775, he moved to London, where he set up an elaborate "Temple of Health" at the newly built Adelphi. At the Temple, health-seeking visitors could enjoy wandering through the ornate, perfumed rooms, listening to the orchestra, or hearing Graham deliver lectures about health, nutrition, and electricity. Sometimes Graham would end his lectures by giving the audience a surprise jolt of electricity, through a wire attached to their seats.

Graham does not seem to have been much of a doctor, but he was a remarkable showman. His lectures were decorated with scantily clad young women, his "goddesses of health," one of whom was reportedly Emma Lyon, who went on to become the notorious Lady Hamilton. He won many fans among high society, including Charles James Fox, and the Duchess of Devonshire, who turned to Graham after failing to conceive.

The centerpiece of Graham's temple was the Celestial Bed, which cost $75 a night, and according to Graham was guaranteed to cure sterility or impotence. The bed was twelve feet (3.7 m) long by nine feet (2.7 m) wide, with powerful magnets beneath, and it could be tilted, to reach the ideal angle for conception. The mattress was stuffed with rose leaves, flowers, and hair from the tails of fine English stallions. A large mirror looked down from the ceiling, and as the orchestra played nearby, perfume (and possibly nitrous oxide) wafted down, and electricity crackled across the headboard, on which were inscribed the words, "Be Fruitful. Multiply and Replenish the Earth."

The Temple was a great success, and soon moved to a new site at Pall Mall. Graham became a well-known public figure, although many dissenters felt that he was a charlatan whose use of sex and titillation were unseemly. However he was also not much of a businessman, as despite its popularity his Temple consistently lost money, and he ended up deep in debt. He eventually returned to Edinburgh, where he began advocating the health-giving properties of mud baths. Presumably, he no longer had the capital to invest in more expensive panaceas.

how did john brinkley make his fortune?

 John Brinkley was a remarkable quack who became one of the richest and most celebrated doctors in America, despite having no medical qualifications. He was born in 1885 into relative poverty in North Carolina, and his childhood ambition was to become a doctor. At the age of twenty-two, he began medical training, albeit at a dubious, unaccredited medical college, but he never completed it, as by this time he had mounting debts, and a young family to support.

In the course of his studies, however, he did come across the concepts of hormones, glandular extracts, and their effects on the body, which would prove to be an important discovery. By 1918, Brinkley was thirty-three, and he had spent more than ten years as a traveling doctor and patent medicine man. He now set up a large-scale clinic in the small town of Milford, Kansas, offering a very unusual treatment, which he marketed aggressively as a miracle cure for impotence, infertility, and other sexual problems. Brinkley claimed he could cure all such ills by surgically implanting goat glands into his patients. The theory was that the vigorous vitality of the randy goat would naturally revive the sexual appetite and capabilities of the human subject.

The procedure was a fairly simple one. A patient would check in to the hospital, having paid a considerable fee of $500 and upward, the equivalent of around $5,000 today. The patient would then be led out to the back of the building, where he or she would select a lusty-looking goat, which would then be castrated. Male (human) patients would then have a slit cut in their scrotum, into which the goat testicles would be inserted, after which the incision would be sutured. Female patients had the goat testes implanted into their abdomen, somewhere vaguely near the ovaries.

This extraordinary treatment made Brinkley a huge fortune over the course of his life, thanks to his skills as a marketer and self-publicist. He was in almost constant conflict with the American Medical Association, who denounced him as a charlatan, but he bought his own radio station, which he filled with advertisements for his own products, clinics, and pharmacies. When this station was shut down by the U.S. authorities, he moved across the border into Mexico, setting up the world's most powerful radio transmitter, allowing him to broadcast into the United States with impunity. It was said that Brinkley's broadcasts could be heard as far away as Finland and the Soviet Union, and the radio signal was so strong that it illuminated car headlights, turned wire fences into receivers, and bled into private phone calls.

Obviously, there was no actual merit to Brinkley's goat gland surgery, and in fact it was extremely dangerous. Many of Brinkley's patients did believe they had been cured, but this was surely only a result of the placebo effect. Many others died, either because of a reaction to the rotting goat testicles that had been implanted inside their body, or because of gangrene or infection caused by Brinkley's unsanitary, unskilled surgical procedures. A recent biography (the excellent *Charlatan* by Pope Brock) suggests that Brinkley, who was well aware that he was a fraud, could be regarded as one of America's most prolific serial killers, with many hundreds of victims.

At the height of his fame, Brinkley was a man

of great standing. He ran for political office a number of times, and was only prevented from winning by the dubious electoral practices employed by his opponents. He was a multimillionaire, with a fleet of cars, yachts, and an airplane, who even loaned one of his yachts to the Duke and Duchess of Windsor. His enormous influence also had some interesting, indirect effects. He filled his radio airtime with local Southern singers and performers, and may have kick-started the boom in bluegrass and country and western music, and established Nashville as its home.

Brinkley's life came to a sad conclusion. He was pursued for decades by Morris Fishbein, the editor of *The Journal of the American Medical Association.* The dispute eventually ended in court, in a case that Brinkley lost, opening him up to hundreds of lawsuits for fraud, malpractice, and wrongful death. By 1941 he was bankrupt. A year later, he was dead, following a series of heart attacks. One obituary described him somewhat euphemistically as "the most fabulously successful medical maverick the country ever saw."

what is a toad-eater?

A toad-eater had one of the most unpleasant jobs in the history of medicine. He (or she) would assist a quack medicine man by swallowing a toad—at the time, toads were believed to be deadly. Many spe-cies do extrude a milky, poisonous substance when threatened. The toad-eater's act was therefore con-sidered to be tantamount to suicide. Having swal-lowed the unfortunate amphibian, the assistant would duly collapse, or perhaps contort his body in agony. The mountebank (so called because he would climb onto a bench to sell his wares) would then pour some of his rejuvenating miracle nos-trum into his assistant's mouth, at which point the poor toad-eater would be miraculously cured. On a good day, this spectacle would be enough to convince several in the audience to part with their cash, and the charlatans could move on to the next town.

Even at the time, it wasn't clear whether or not toad-eaters did actually eat the toad. There are lots of records of people having heard of someone who had once seen it happen, but no first-hand accounts. Presumably, the toad-eater would often palm the creature, and simply pretend to eat it. Alternatively, they may have used a nonpoisonous frog. Perhaps sometimes they did swallow the toad, and simply accepted the resulting few days of ill-ness as the cost of doing business. We don't know, and at present scientists seem to be focusing their research on other priorities, more's the pity.

The practice of toad-eating is now largely forgot-ten, but it has had one enduring influence, as it gave us the term "toady." A toad-eater was someone who would risk illness and even death for his boss, which suggests he would have been an unusually lowly, supine, obsequious type of person. Thus, a toady today is a sycophantic employee who will suck up to his or her boss, and suffer any humiliation.

who was the münchhausen of münchhausen syndrome?

The answer is Karl Friedrich Hieronymus, Freiherr von Münchhausen, a German baron who was born in Bodenwerder in 1720. Münchhausen did not suffer from this syndrome, or diagnose it; rather, he became famous for telling tall tales, and thus when the condition was discovered, more than a century after Münchhausen's death, it was named after him in tribute. Münchhausen syndrome is a psychiatric disorder in which the patient feigns illness in order to gain attention and sympathy. There is a related condition called Münchhausen syndrome by proxy, in which a parent pretends that his or her child is sick, or even induces sickness in the child, to gain attention for himself or herself.

Baron von Münchhausen is believed to have told numerous astonishing stories about his adventures. He served in the Russian army, including two

campaigns against the Ottoman Turks, before re-
turning home to Germany. There, he is said to have
recounted astonishing stories about his adventures
abroad, including claims that he had traveled to
the moon (and met the moon people), fought with
bears, flown on a cannonball, and visited an island
made of cheese.

His amazing stories were published in 1785 by
Rudolf Erich Raspe, with the excellent title *Baron
Münchhausen's Narrative of His Marvellous Travels
and Campaigns in Russia*. Many of the book's tall
tales are actually traditional folk stories that were
around before the Baron, and so it is thus not clear
whether this collection was intended to celebrate
the Baron, or to damage his reputation. Whatever
the truth of the real Baron's life, this collection
made his name synonymous with fantastic, imagi-
native fantasies, and so when Richard Asher first
described a syndrome in which patients would
invent and simulate illness, he "respectfully" dedi-
cated it to the Baron.

does organ theft actually happen?

There is an enduring urban legend about a gang
of civilized organ thieves, a legend which has been
circulating for perhaps twenty years or more, presumably as a kind of reaction to the real-life practice

of organ donation becoming more and more common. The story has been passed around e-mail and
the Internet in many different incarnations, and
has made its way into newspapers, TV dramas, and
horror films.

The story generally goes something like this: A
lone traveler is approached by a glamorous stranger,
usually while having a drink in an airport lounge.
The next thing he knows, he wakes up in a hotel
room, in a bathtub filled with ice. On the mirror are
written the words, CALL 911. DO NOT MOVE, and
there is a phone by the side of the bath. When the
ambulance crew arrives, they tell him that his kidney
has been removed by an organized gang who steal
and sell organs on the black market.

This version of the story is clearly false. There are
no reports of any kidneys being stolen in the United
States, and many of the details are clearly implau-
sible. For one thing, a kidney cannot be implanted
into just anyone, there needs to be a precise blood
and tissue match between donor and recipient.
Second, if an organ is to be transplanted, it usually
needs to go into the recipient very quickly if it is
to be viable, and they may even need to be present. Third, this kind of operation is complex and
demanding, usually requiring about five people to
perform it, over a period of hours, in a sterile operating room, using various pieces of heavy medical
equipment, of the type which couldn't be easily
smuggled into a hotel room.

The fact that this version of the story is clearly an urban legend doesn't mean that organ theft never happens. There is a real, commercial market for organs, with organ brokers charging as much as $150,000 for arranging and managing a transplant. Willing donors in the Third World will sell a kidney for as little as $3,000, perhaps even less—this is more than a year's wage in many parts of the world, and people can usually function perfectly well with only one kidney. Demand for organs consistently outstrips supply—at any given time, there are around 100,000 people in the U.S. awaiting lifesaving organ transplants, but because of a shortage of available organs, fewer than 30,000 transplants take place each year. On average, 18 people die each day because of a lack of available organs for transplant. People who need an organ are therefore highly motivated.

Thus, it is not entirely surprising to find that it does happen. There are no recorded instances in the United States, at least not yet, but there have been reports from other parts of the world, in particular India, which used to have a successful legal organ trade before legislation was passed in 1994. In January 2008, several people were arrested in the Indian city of Gurgaon for running an organized, large-scale kidney theft ring. According to reports, hundreds of men were lured to an underground facility where they were promised jobs, but then tricked or forced into having a kidney removed, to be transplanted into a wealthy recipient. At the time he was arrested, the alleged ringleader Amit Kumar was said to be trying to flee to Canada.

So, the lesson seems to be that you don't need to worry about organ gangs trying to drug you in an airport lounge, but if you're offered mysterious, well-paid, cash-in-hand work in an underground medical facility in India, by men brandishing scalpels, you may want to think twice.